Dictionary of
FILM and TELEVISION TERMS

For Jerry who
loves movies!

Juliette

10/19/85

Other books by Virginia Oakey

Screenwriter's Handbook (with Constance Nash)
Television Writer's Handbook (with Constance Nash)

Dictionary of FILM and TELEVISION TERMS

Virginia Oakey

BARNES & NOBLE BOOKS
A DIVISION OF HARPER & ROW, PUBLISHERS
New York, Cambridge, Philadelphia,
San Francisco, London, Mexico City, São Paulo, Sydney

 For information address Harper & Row, Publishers, Inc., 10 East 53rd Street, New York, N.Y. 10022. Published simultaneously in Canada by Fitzhenry & Whiteside Limited, Toronto.

FIRST EDITION

Library of Congress Cataloging in Publication Data

Oakey, Virginia.
Dictionary of film and television terms.
1. Moving-pictures—Dictionaries. 2. Television broadcasting—Dictionaries. I. Title.
PN1993.45.034 1983 384'.8'03 82-48254
ISBN 0-06-463566-X

---

83 84 85 86 87 10 9 8 7 6 5 4 3 2 1

Acknowledgments

—My warm thanks to Clem Portman, writer and poet, whose contributions drawn from his long and successful career in the film industry have been invaluable in the compilation of this dictionary.

—My gratitude to Gay Foster for her tireless assistance in my efforts to achieve validity and accuracy in these definitions; to Isabelle Ziegler, writer-editor, for her indefatigable editorial assistance; and to Nancy Cone, my editor at Barnes & Noble Books, for her sustaining interest and remarkable editorial skills.

—My appreciation for research materials from the following libraries: The American Film Institute in Los Angeles, the University of California at Irvine, Saddleback College at Mission Viejo, California, and the public library at San Juan Capistrano, California.

Note to the Reader

Words printed in SMALL CAPITAL letters are defined elsewhere in the dictionary.

Dictionary of
FILM and TELEVISION TERMS

A

AAAA (American Association of Advertising Agencies). An organization that exchanges information and decrees general policy and standards for the industry.

AAAA (Associated Actors and Artistes of America). An international organization composed of national trade unions for persons employed in the field of performing arts.

A and B printing, AB printing, A and B roll printing. Printing from original (usually positive) film that has been conformed into two rolls, each of which has alternating shots and blank opaque LEADER. A and B printing makes possible the elimination of splice marks in release prints. It also allows the printing of dissolves and supered (superimposed) titles.

A and B rolls, AB rolls. Two or more matching rolls of film that make duplication possible by having alternate scenes intercut with opaque LEADERS. That is, beginning at a single point, roll A presents a picture to the duplicate being printed where roll B presents an opaque leader, and vice versa. This allows double or multiple exposures in printing to obtain title or picture superimpositions and dissolves from one scene to another. It also provides a checkerboarding to prevent the appearance of film splices on the screen.

A and B roll supered title. Titles that are superimposed over the action seen on the screen, achieved by double printing A roll over B roll and vice versa.

A and B winds. The location of EMULSION applied on either side of 16mm SINGLE-PERF film base. "A" wind has the emulsion applied toward the reel hub and is usually used for contact printing. "B" wind has the base toward the reel hub and is used for camera raw stock, projection printing, and optical work.

ABC (American Broadcasting Company). One of the three major American television networks.

aberration. Image distortion created by signal interference or the misalignment of the electron beam. Malfunction of an optical element such as lens, prism, or mirror that creates a systematic distortion. (Aberrations in early lenses included curvature of field, distortion, and spherical and chromatic differences.)

abrasion marks. Imperfections on film created by external substances such as dust, grit, grease, and emulsion pile-ups.

absorption. The means by which transmitted light is retained within a lens.

absorption filter. A light filter that is effective in obstructing certain wave lengths of light and transmitting others.

abstract film. **1.** A film that shows only the essential elements (shapes, patterns, etc.) of the original subjects. **2.** Any nonrepresentational film.

AC adapter. Step-down transformer that converts battery-powered equipment to electric power source.

Academy aperture. Standard governing film framing established by the American Academy of Motion Picture Arts and Sciences. Controls the actual size of the frame mask in 35mm cameras and projectors.

Academy Awards. Awards presented annually by the ACADEMY OF MOTION PICTURE ARTS AND SCIENCES for artistic achievement in several categories, the winners being chosen from five nominees in each field.

Academy of Motion Picture Arts and Sciences. An organization of motion-picture producers, directors, actors, and technicians. Presents the ACADEMY AWARDS (OSCARS) in an annual ceremony.

Academy of Television Arts and Sciences. A Hollywood-based organization of professionals in the television industry; sponsors the annual Student Film Awards and EMMY awards for outstanding achievements.

Academy standards. Standards governing projectors, film leaders, and camera apertures, established by the Academy of Motion Picture Arts and Sciences.

accelerated motion. Action that is filmed while the camera rate is progressively reduced. Frequently used to represent action as taking place at a greater than normal speed.

acceptance. Local affiliate station acceptance of network program.

acceptance angle. Two-dimensional angle received by a lens or light meter.

access. Availability to public of cable production time. See PUBLIC ACCESS.

access time. Time during playback of videotape between the moment information is asked for and the moment it is transmitted.

account. Advertising sponsor.

account executive. Member of the advertising agency who serves as liaison between TV channel and the sponsor.

account group. Creative or management personnel of an advertising agency assigned to one particular client.

ace. A spotlight with a 1,000-watt lamp or bulb.

ACE (American Cinema Editors). An organization of professional motion-picture and television film editors; presents annual Eddie Award for best film editing.

"A" certificate (Adult Certificate). A designation given by the British Board of Film Censors to a motion picture that may be seen by children under sixteen only when accompanied by parent or guardian.

acetate. 1. Slow-burning base material used in motion-picture films. **2.** Transparent plastic sheet form used as art work surface in animation work. **3.** Separately cut (not pressed) phonograph disk composed of aluminum coated with cellulose nitrate.

acetic acid. One of the properties of film cement, made of a colorless liquid solvent.

acetone. Organic liquid solvent used in the production of certain kinds of film cement and to clean film-splicing equipment.

achromatic. 1. Designates any optical product that has been corrected for longitudinal and/or lateral chromatic aberration or imperfection (*apochromatic* designates an even more meticulous chromatic correction). **2.** Without color.

acoustic. Early nonelectronic process used for disk recording.

acoustical feedback. Combined reaction between the microphone and loudspeaker in a public-address or other sound system, which results in a rasping-screeching sound.

acoustics. 1. Science dedicated to the study of sound. **2.** The attributes of a room, sound-recording stage, auditorium, or other enclosure as they affect the resonance qualities of sound.

across the board. Referring to television programs or other broadcast materials scheduled at the same time each week. Cf. ONE-SHOT, SPECIAL.

ACT (Action for Children). A voluntary organization of citizens concerned with upgrading the quality of television programs for children.

actinic light. Visible or ultraviolet light that provokes chemical or electrochemical action.

acting. Performance of a role in a film or other dramatic event.

action. 1. Subject's motion within camera range, to be incorporated in film being shot; usually rehearsed, occasionally spontaneous. **2.** Director's command indicating that actors are to begin performance.

action cutting. The instantaneous shifting from one shot to another, which is designed to give the impression that the action is uninterrupted, despite obvious changes of camera position. Achieved by overlapping the action on successive shots so that the first shot seems to include the beginning of the second shot or by having two cameras in simultaneous operation so that the transfer from the first shot to the second seems to be continuous action. The technique is also used to edit shots in which performers enter or leave the camera action range, and shots in which action starts or stops within the same frame.

action field. (Frequently referred to as FRAME or SHOT.) The area that is actually being filmed by the camera.

action properties. (Often referred to as action PROPS.) Things actually used (or handled) by performers in a film; e.g., a table clock is merely a prop until a performer picks it up, throws it, or touches it in some bit of action, making it an *action prop.*

action still. A still (unmoving) shot of an action scene as it appears in the final or theater version of the motion picture.

action theme. Condensed and concise version of the major action of a motion picture.

actor. Any person, male or female, who performs in a film.

Actors' Equity Association. A trade union for actors.

AC transfer. Videotape duplication by contact in high-frequency field between high-coercivity master and low-coercivity slave. See SLAVE.

ACTT (Association of Cinematograph, Television and Allied Technicians). British film and television union representing members of these technical trades.

actuality. Film of an actual event as it happens.

actual sound. Sound that has a visible or implied source on the screen during filmed action; e.g., dialog by visible actors or one whose presence has been previously established; the ticking of a grandfather clock that is seen or has been shown to be present.

acutance. A measure of the sharpness of an edge in a filmed object.

AD. See ASSISTANT DIRECTOR.

adapt. 1. Means by which one camera component is attached to another component, though not ordinarily used in this manner. **2.** To write a film script from material based on another source, usually a novel, short story, or play, but occasionally a biography, autobiography, or a newspaper or magazine article.

adaptation. Rewriting of fact or fiction for presentation in a film, usually as a complete script or a treatment to be developed in a film production.

adapter. A device, such as plug, jack, or outlet, used to connect an additional element or an attachment of dissimilar size to a piece of equipment.

ADC (analog-to-digital converter). Equipment designed to dissect analog television signals into digital transmission form.

added scenes. Material, shots, or sequences written into a script during its filming or after its completion.

additional dialog. Actors' words added to script while film is being shot or after its completion.

additive primaries. Basic television colors, red orange, green, and blue violet. Used in different combinations to produce all other colors and white.

additive printer, additive color printer. Equipment that prints from color originals or intermediates and employs light of three colors (blue, green, and red) that are individually controlled to present a composite color printing light at the point of exposure of the print stock.

additive process, additive color process. A photographic means of viewing and rearranging colors of a particular scene with three light filters, each of which represents a primary color. The process involves the shooting of a black and white negative of the subject through each filter, then making a black and white positive transparency from each negative. The three positives are projected simultaneously on a screen, each through a specific filter used in making the negative from which it originated. The superimposed images then appear in color.

addressable. See PAY-PER-VIEW.

address code. Digital videotape retrieval system that uses control or cue track signals. Occasionally called *birthmark.*

ADI (area of dominant [station] influence). A classification of the American Research Bureau market research service that designates the area groups in which most television viewers watch their local channels.

adjacencies. Material—audio or visual or both—that precedes or follows a specified television program or commercial.

ad lib. (From *ad libitum*, Latin for "at pleasure.") Improvised material, usually vocal. Spontaneous response without rehearsal.

Adult Film Association of America. An organization of producers, distributors, and exhibitors of hardcore, sexually explicit motion pictures.

advance. Number of frames between the picture and synchronous sound on composite film print; adjusted to the projection pull-up requirements. There are about twenty frames in 35mm, about twenty-six frames in 16mm film.

advertising agency. Independent company that prepares advertising for clients, usually not in competition with other clients.

Advertising Council. Group that sponsors public-service material with the assistance of various advertising agencies.

advertising director or manager. Executive responsible for the planning and supervision of advertising; serves as liaison with agency.

advisory. Inside information about forthcoming stories presented in news service "bulletins."

aerial, antenna. Special type of conductive device that radiates or receives audio frequency broadcast signals. Cf. SATELLITE.

aerial cinematography. The shooting of films from cameras mounted on airborne vehicles such as airplanes, helicopters, or hot air balloons.

aerial image. True image reproduction formed at a plane in space by an especially designed optical system.

aerial image animation stand. A specially constructed stand that uses a projection system to reproduce an aerial image at the platen plane, allowing the superimposition of cell animation over live action or other images.

aerial mount. Device used to attach cameras to airplanes or other airborne vehicles.

aerial perspective, atmospheric perspective. Optical illusion that makes distant objects appear to be lighter in color and smaller in form than they really are. It may be created by natural visual reaction to things viewed from a distance or contrived by back lighting or by side lighting.

aerial shot. Camera shot made from airborne vessel. On television these are frequently stock shots.

AFC. See AUTOMATIC FREQUENCY CONTROL.

affiliate. A broadcast channel under contract to a network and required to use more than ten hours a week of network programming.

AFI (American Film Institute). A nonprofit corporation that offers two educational programs: **1.** a one-year Curriculum Program open to filmmakers with some proficiency in their crafts (also considered are individuals with experience in related fields such as literature, music, the theater, photography, or the fine arts); **2.** a Conservatory Program in which emphasis is placed on the work of the individual in his specialized field. Members of this program are selected from among Fellows who have completed the Curriculum Program. The institute also has an endowed library that is open to serious movie buffs and aspiring film writers.

AFM (American Federation of Musicians). A union whose membership comprises musicians and music performers (vocalists).

AFTRA (American Federation of Television and Radio Artists). A union whose membership comprises radio and television actors, singers, and creative sound-effects specialists.

AGC (automatic gain control). A servo circuitry that offers consistent signal levels.

agency. See ADVERTISING AGENCY; TALENT AGENCY; LITERARY (OR WRITERS') AGENCY.

agency of record. The agency placing television and radio advertising that was prepared for a single corporate advertiser.

agent. A person who represents an artist (actor, director, writer) for a percentage of his/her salary or occasionally for a flat fee. The agent represents the client's interests in contracts, sales, and other negotiations with film production companies.

agitation. Turbulence deliberately created in film-processing BATHS.

AGVA (American Guild of Variety Artists). A union whose membership comprises performers.

aided recall interview. In-home audience survey method that uses "clues" to judge viewing and listening habits of home audiences.

air date. The day on which a specific program will be broadcast.

air gap. The narrow gap between the two elements of a magnetic recording or playback head.

air quality. Program material that is produced according to technical broadcast standards.

air squeeze. A device used to spray a continuous stream of air on motion-picture film as it moves from the final wash to the DRY BOX, in order to blow off the water.

airwaves. 1. The medium of radio or television broadcasting. **2.** The broadcasting itself. ("The airwaves were dominated by news of the assassination.")

Akai. Japanese electronics firm that manufactures hand-held cameras so vital to television filming with its need for mobility.

alignment. Desired electronic balance.

Alleflex machine. Registered name for equipment used in silent film era to produce sound effects.

Allen screw. Special flush-mounted machine screw with a hexagonally shaped insert head.

Alliance of Motion Picture and Television Producers. An organization of studio producers recently formed (1982) to deal effectively with labor problems that occur within the industry.

alligator. A temporary circuit clip attachment with sawtooth edges resembling the jaws of an alligator.

allocation. The licensing of frequency and power by the FCC to certain broadcast stations.

Ally Pally. The Alexandra Palace, the location of the British Broadcasting Company's first television productions.

alpha wrap. Videotape wind configuration around a helical scan drum.

alternate sponsorship. Manner in which large and small sponsorships are rotated in a broadcast series in order to reduce the cost of advertising.

alternative television. Television programming provided by sources other than the establishment or network-sponsored or -produced material.

ambient light. Lighting, of a set or scene, which does not fall directly on the subject of the shot. May be placed above or behind the subject.

ambient temperature. Temperature of the gas or liquid that surrounds or encompasses equipment.

American National Standards Institute (ANSI). The nonprofit organization dedicated to the establishment of technical standards for film and television on an international level. (Formerly the American Standards Association.)

American Television and Radio Commercials Festival. An event held annually to honor winners of the broadcast commercials competition.

American Women in Radio and Television. An organization of professional women in creative, executive, or administrative positions in local television stations, television networks, radio, and advertising agencies.

ampere (Amp). A basic unit of electrical current.

Ampex. Registered name for a process that electronically records a television picture and its sound onto magnetic tape.

amplifier. An electronic contrivance that makes an electronic signal louder without diminishing its power; a mechanism that increases the power or voltage of an incoming signal.

amyl acetate. A liquid used as a component of film cement.

ANA (Association of National Advertisers). A corporate group that disseminates information among its members and decides general policies and industry standards.

anaglyph. The combination of two pictures that gives a three-dimensional effect when the superimposed images on the screen are viewed through an *anaglyphoscope* that has two eyepieces of different but complementary colors matching those of the picture.

anamorphic image. An image that is squeezed in a single direction through unequal magnification by an ANAMORPHIC LENS.

anamorphic lens. A lens whose purpose is to achieve a wide-screen image by projecting a picture through a correcting lens. It is designed to distort an image by means of elements having cylindrical surfaces. The lens compresses the image along the axis of the focal plane. At 90 degrees from that axis, the image remains unaffected in the focal plane direction. Thus, wide-screen pictures can be projected by the use of standard-width camera film.

anamorphic release print. A film print in which the images are compressed horizontally.

anastigmat lens. Lens that are corrected optically for deviations in horizontal and vertical planes.

anchorman/anchorperson. On television news broadcasts, the major figure and coordinator.

anechoic chamber. A soundproofed room used to test acoustical properties.

Angenieux. Trade name of a French lens system.

angle. See CAMERA ANGLE.

angle of acceptance. The portion of action covered by the camera, both horizontally and vertically.

angle on. Direction written in a film script indicating that another camera angle is to be made of a previous shot. Usually used to emphasize a specific object in the shot, such as ANGLE ON GIRL LOOKING OVER FENCE. (The previous shot may have included not only the girl and the fence but a wider view in which a house and a road are also visible.)

angle shot. A shot that continues the action of a preceding shot but from a different direction or angle.

angstrom unit. A unit of length equal to one ten-millionth of a millimeter, used to express electromagnetic wave lengths.

animation. The cinematic process of creating the illusion of mobility from inanimate subjects or drawings. This is accomplished by exposing film in one, two, or three frames followed by slight progressive changes in the subjects being filmed.

animation board. A device, usually a studded drawing board, used to prepare artwork for animation.

animation camera. A film camera mounted above the subject bench or board used to make single-frame exposures.

animation cameraman. Any cinematographer who is skilled in the operation of equipment especially designed for animation films.

animation crane. See ANIMATION STAND.

animation designer. Cartoonist who creates master drawings for an animation series.

animation stand. Support for the camera that makes it possible to raise or lower the camera above the peg board or animation bench.

animator. Cartoon artist who draws the figures for the master sequences sketched by the designer.

announce booth. A small soundproofed room on set or stage in which individual voice recordings are made.

announcement. A commercial. "We'll be back after the following announcement."

announcer. Person who introduces a television program, gives commercials, and identifies stations.

answer print. The first completed color-and-sound print ready for release. If it fails to meet the producer's standards for subsequent prints, it is reevaluated for further editing and/or other changes.

antagonist. The villain of a film, the actor/actress who epitomizes evil or brutality, without whom the conflict that engages the hero would be absent. Occasionally the antagonist is a force of nature, such as heat wave, typhoon, or hurricane, against which the hero must battle for survival.

antenna array. Several elements in a system that receives or sends signals.

antenna farm. A cluster of station antennae arranged to avert aerial navigation accidents.

anthology film. 1. A full-length (120 minutes or more) feature film that presents excerpts from other feature films. **2.** A collection of short films.

anti-abrasion coating. A thin film applied by manufacturers to the base side of certain films in order to eliminate or diminish scratching or other abrasive damage.

anticlimax. Anything that occurs in the final minutes of a film that causes audience letdown.

antihalation backing, antihalation coating. An opaque film coating on the base side of film that is an antireflexive agent. See HALATION.

antihalation dye. A dye added to the film base to further reduce HALATION.

antihero. The protagonist, male or female, in a film, who has pronounced personality or character defects or eccentricities not usually associated with heroes.

AOC, Ass On Curb. Assistant director's order to crew members concerning transportation by studio bus: e.g., "6:00 A.M., AOC, Fifth Street and Vine."

aperture. 1. Lens aperture: the orifice at the front of a camera that admits and limits the amount of light or electrons that pass through the lens to a negative film. **2.** Camera aperture: in 16mm moving-picture cameras, the opening is 0.410 × 0.294 inches, which controls the area of each exposed frame. **3.** Projector aperture: in 16mm projectors the opening is 0.380 × 0.284 inches, which controls the area of each projected frame. **4.** Printer aperture: the opening through which light passes to expose the film.

aperture plate. The plate in a camera or projector in which a controlled orifice limits the action area that is recorded on film or projected on the screen.

"A" picture. Feature film in a theater, to be shown first on a double bill, or simply a highly touted and budgeted film.

APO (action print only). An optical print negative without a sound track.

apochromatic lens. A lens that has been corrected by the manufacturer for any spherical and chromatic deviations.

apple, baker, double apple. Terms spoken to designate SLATE numbers that have been given alphabetical initials in order to assist the editor in listening to the sound track: apple for the letter A, baker for B, double apple for D, etc.

apple box. A sturdy wooden box or half-box used to achieve a desired position for an actor or object in a camera shot. Also used to increase the height of an actor or object.

approach. The director's instruction to bring the camera closer to the object of the shot.

appropriation. Television term applying to the approved estimated cost of an advertising campaign.

apron. Extension of a stage that stretches beyond the proscenium arch.

ARB (American Research Bureau). A television audience market research survey distributed three times a year that is based on the results of questionnaires on viewing habits.

arc. The movement of a camera DOLLY along a curved path.

archetype. A film that is typical of a certain genre in style, content, and presentation.

arc light (lamp). A lighting instrument used for projectors or studio lights in which the source of illumination is the electric discharge across the gap between two carbon arc poles.

arc out. The director's instruction to a DOLLY PUSHER to move the dolly away from the scene along a curved path.

A-roll original. The original film that has been rolled onto a single reel for printing.

Arri. The abbreviation for the Arriflex camera (from the names of the inventors, "Ar" for Arnold and "ri" for Richter). A lightweight reflex moving-picture camera, in both 16mm and 35mm speeds.

art card. A black cardboard on which type or designs for film photography are hot-pressed in white.

art department. The personnel responsible for preparing graphic materials needed for film productions.

art director. The supervisor responsible for the conception and design of all sets, including decor, used in a film. Frequently scouts and chooses the natural locations used in the film.

art house. 1. A motion-picture theater that presents foreign or experimental films. **2.** A theater that generally shows revivals of classic films.

art still. 1. A painting to be used as a slide for projection at the back of a scene or as a prop. **2.** A posed photograph of an actor.

artificial light. Light produced in any manner—from candles to electricity—as opposed to natural light from the sun or sky.

ASA exposure index, ASA, ASA speed. Letters that refer to the numerical-exposure index of a film according to designations stipulated by the American National Standards Institute.

American Society of Cinematographers (ASC). An honorary society of motion-picture cameramen/women, based in Hollywood.

analog image synthesis. A method of producing computer graphics used in special effects in which images can be made to spin, oscillate, shrink, and achieve other special visual effects in fantasy films.

ASCAP (American Society of Composers, Authors, and Publishers). A trade guild that protects the performance and publication rights of the three groups.

ascertainment. FCC stipulation governing the licensing of broadcast stations that requires them to be responsive to local programming needs.

ashcan. A 1,000-watt floodlight.

aspect ratio. The relationship, expressed in width-to-height ratio, of a projected film picture or shot to the screen that it fills.

assemble. In film editing, to complete the first process of organizing and joining the filmed shots in approximately the order or sequence required for the completed version; to produce a rough cut.

assembly. Film shots that have been printed and spliced together in correct sequence according to the script.

assistant cameraman. The person who aids the cameraman or director of cinematography, particularly in checking the equipment, loading film, and following focus. Seldom, if ever, operates the camera.

assistant director (AD). The person who performs many and various functions assigned to him by the director, including scheduling, supervising crews, conducting rehearsals or action shots, and generally acting as the "detail man."

associate editing, associative editing. The alignment of film shots in proper order so that contrasts, comparisons, and similarities may be studied.

Association of Motion Picture and Television Producers (AMPTP). An organization of the principal motion picture and television producing companies; affiliated with the MOTION PICTURE ASSOCIATION OF AMERICA.

associate producer. 1. Person to whom the head producer of a large studio assigns the production of a certain film or films. **1.** Assistant to the producer.

astigmatism. In film terms, the lens design that causes light rays passing through the lens to converge imperfectly.

asynchronous sound. Sound that is part of, or resulting from, the filmed action but is not precisely synchronized with it. In some instances, the source of the sound may be OFF-SCREEN (OS) but is assumed to be present.

atmosphere. The tone or dimension added to the action by visible or invisible qualities or elements such as rain, heat, danger, tranquility.

ATS (automatic transmission system). Monitors and adjusts itself, requiring little, if any, engineering supervision.

attenuate. To decrease signal power.

attenuation loss. The reduction of the quantity of electrical energy in an electrical cable or sound wave.

attenuator. A device that diminishes the signal amplitude in a cable.

audience. Persons in a movie theater or viewers of television who are watching a filmed production. Broadly, all persons who attend movies or watch television.

audience accumulation. The overall audience, accumulated through repeated exposure, as reported by a research survey tabulation.

audience flow. The pattern in which some viewers stay tuned to the same television channel, some change to another channel, and some turn the set off, as reported by a half-hour research count.

audience rating. Any judgment, verbal or written, made by an audience in the appraisal of a film.

audience share. The percentage of local or national households with one or more television sets that are turned on during a particular program.

Audimeter. An electronic device used in the 1,250 NIELSEN sample homes that record television viewing habits of that family.

audio. 1. Of or pertaining to the transmission or reception of sound. **2.** Sound that includes dialog, music, narration, and special sound effects. **3.** All equipment and components used to reproduce sound, such as circuits in a receiver. **4.** Frequencies or signals within the audible range.

audio frequency. Usually an audible sound wave (between 15 and 20,000 Hz). Standard audio frequency ranges are bass (0–60 Hz), mid-bass (60–240 Hz), mid-range (240–1,000 Hz), mid-treble (1,000–3,500 Hz), and treble (3,500–10,000 Hz). See HERTZ.

audio mix. Two or more sounds combined electronically into a single SOUND TRACK.

audion. An electronic amplifier tube invented by De Forest, 1906.

audiotape. A plastic tape with a coating of magnetizable metallic oxide on which sound can be recorded magnetically. Available in widths of one-eighth, one-quarter, one-half, one, and two inches on cores up to 7,200 feet.

auditions. A test performance, given before an employer of theatrical talent, given by actors and musicians (and those in related fields, such as mimes, magicians, and acrobats) who are seeking employment.

auteur. (The French word for author.) A term used in film criticism. In this sense the *auteur* director has artistic control of his film so that it carries his individual imprint.

automatic frequency control. The automatic control of an OSCILLATOR in a television transmitter or receiver to reduce unwanted frequency changes.

avant garde. An experimental film that may be a forerunner of a new genre.

B

baby, the baby. Nickname for the film camera.

baby legs. Short-legged tripod used for camera work.

baby plate, baby wall plate. A metal plate with a short upright pole used for mounting a small spotlight.

baby spot. Small spotlight with a 500–750-watt lamp used in studio or set lighting.

baby tripod. A low tripod used on table tops or for low-angle camera work.

back. To add musical accompaniment.

backdrop. An artificial background, usually painted on a CYCLORAMA, curtain, or flats, used to achieve the effect of a natural setting such as forest, beach, or other landscape in a shot or sequence.

back focus. The distance from the focal plane to the back of a lens set at infinity.

background. 1. The space in the action area of a shot or shots that is farthest from the camera; any area that is behind the subject being photographed in the foreground. **2.** Music played to underline action being filmed. (See BACKGROUND MUSIC.)

background action. Action that takes place behind the subject of the shot in the foreground.

background film. An educational film that provides information on a specific subject.

background light. Any artificial light used to illuminate the background of a shot or sequence.

background music. Music during a film, usually on a sound track but occasionally by live performance, played by musicians or heard from records or tapes. It serves to emphasize the action or to form a (usually) unintrusive accompaniment to monologs or dialogs.

background noises. Minor sounds used to add realism to a scene (as the clatter of dishes in a restaurant scene).

background plate. A glass slide on which a photograph appears in a back-projection system.

background projection. The projection of immobile or mobile images on a translucent screen before which action or film titles are photographed. Often used to provide moving action for actors in the foreground on the set or where normal front projection is not feasible.

backing. 1. The coating on the back of film intended to diminish abrasive sounds. **2.** A flat background, usually a large photograph or painting on a plain surface, placed behind actors in an action shot.

backing removal. A procedure of processing film stocks with ANTIHALATION backing, to chemically soften them in preparation for buffing.

backlash. Undesirable play in mechanical connections due to maladjustments.

back lighting. To light the subjects of a shot from the rear.

back lot. A large outdoor area of a major film studio equipped with exterior sets that contain streets, false-front buildings, and other property used to simulate actual locations.

backpack. Portable equipment that may be carried on the back for recording camera-signal transmissions.

back projection. Projection of film on a translucent screen from a rear projector to provide what appears to be a moving background for actors being filmed on a set.

backstage. Any area behind the scene of action on the set that does not appear on the screen.

backtime. The backward synchronizations of program components.

back-to-back. Referring to program components or entire programs that are run consecutively.

back-up interlock system. Magnetic sound playback sets, used with a pick-up recorder, that can be reversed through erasure to create better sound during a mix.

back-up schedule. A schedule of shots, called for in the script, that can be substituted in case something, such as an actor's illness, bad weather, poor outdoor lighting, or mechanical or technical problems, interrupt the planned shooting schedule.

baffle. 1. Mobile acoustical wall or batting near a sound stage, used to control reverberation during recording sessions. **2.** A part of a sound-reproducing system that is in the loudspeaker unit itself or in its cabinet.

balance. 1. Juxtaposition of instruments and microphones to achieve the best sound in a recording session. **2.** Rearrangement of the frequency response of an audio circuit to overcome any change in the frequency responses of the signal.

balance stripe. The narrow strip of magnetic coating applied to the edge of the film opposite the magnetic sound track that flattens the film stock as it winds over magnetic heads.

ballast. A device that governs the supply of electric current as it travels from one source to another.

balloon. See AERIAL.

balop. A large slide. See BALOPTICON.

balopticon. The trade name of a camera device that transmits four-by-five-inch opaque art cards. Named for innovators Bausch and Lomb, using the first two initials of each name.

banana plug. A testing connector.

banding. A speed distortion or irregularity in a videotape playback head.

bandpass. Circuit removal of unwanted frequencies.

bandshaping. The process by which Q and I signal BANDWIDTHS are decreased in order to fit designated color transmission.

bandwidth. The number of radio-frequency-modulated signal frequencies in a specified channel: 6 MHz.

bank. 1. A group of lights or other lighting equipment. **2.** To make such a group ("Bank the lights above the action.").

bankroll. To supply the funds for the production of a film ("Mr. X will bankroll the film.").

BAPSA (Broadcast Advertising Producers Society of America). An organization of commercial producers.

BAR (Broadcast Advertisers Report). The report of an advertising research service that oversees commercial use of the network or the market.

bar sheet. A chart depicting recorded dialog and the number of frames of duration for each syllable and pause, principally used in animation films.

barn door. Side or top flaps or flippers on hinges, attached to a lighting device, that can be adjusted to control the light; often used to create desired shadows or to block out a microphone shadow.

barn-door wipe. On-screen imitation of doors opening.

barney. A cover for a camera designed to protect it from rain, snow, blowing sand, excessive heat, etc.

barrel distortion. Lens distortion that makes the sides of square objects appear to bulge outward.

barrel mount. A device holding the lens elements that is attached to and extends from the camera.

barrel shutter. A cylindrical shutter that can be rotated for use on projectors that have two opposite APERTURES.

base. 1. Film substrate EMULSION coating that is sensitive to light. **2.** Plastic audiotape or videotape substrate with magnetizable metallic oxide coating. **3.** Cosmetic foundation for makeup.

base density. The density after fixation in the portion of film that remains unexposed in a negative film.

base down, base up. 1. Terms applied to correct threading of various generations of film through editing equipment. **2.** Terms applied to the usual positions of certain lamps.

base light. Diffused or general lighting arranged to raise the level of lighting on a set before modeling lights are put in place.

base plate. A pedestal set on a perpendicular shaft to serve as a mounting post for lighting equipment.

base side. The side of a film opposite the EMULSION side.

base-to-emulsion. A particular winding of film to be placed in a printer in order to change the ordinary GENEALOGY of the film. The print would have its EMULSION and SOUND TRACK in B-wind position. See B-WIND.

base-to-base splice. A splice made with the base side of the end of a section of film overlapping the base side of another piece.

basher. 500-watt circular floodlight.

basic network. A specific collection of affiliate channels offered by a network for advertising commitment on a national level.

bass boost. Low sound frequencies intensified by use of an electronic device or component.

bass rolloff. The reduction in the low-frequency energy released in a specific sound.

bat blacks, bat down. A term meaning to adjust the black tones in a television picture.

batch. A mixture of chemicals.

bath. 1. A liquid chemical solution used in processing film; also a water rinse. **2.** The tank used in a laboratory for developing film.

batten. A horizontally suspended pipe from which lights or scenery can be hung.

battery. A case in which DC electric power is stored.

battery belt. A plastic or leather belt on which the operator of a hand-held camera suspends rechargeable power cells that have connections for camera power cables.

battery light. A portable battery-operated light.

bay. 1. Area in a studio used for storage of set properties. **2.** Racks on which equipment is mounted.

bayonet mount. A mount into which camera lenses are inserted with a snap-lock device requiring no threads.

bazooka. A support used to hold an overhead LUMINAIRE with its various components.

BBC (British Broadcasting Corporation.) A noncommercial independent organization that provides radio and television services, operating by a royal charter since 1927 under a license and agreement with the postmaster general.

BCU (big close-up). Shows less than an actor's entire face. Written in a script, it would read, "BCU on diamond earring worn by actress on her left ear." The camera is, or appears to be, closer to the subject than is usual for a close-up.

beaded screen. A highly reflective front-projection screen covered with small glass or plastic beads.

beam. A tiny electron stream directed through cathode guns.

beam angle. A light projecting a beam that has half the energy output of a spotlight.

beam lumens. The quantity of light within the beam angle.

beam projector. A spotlight that projects a thin light beam.

beam splitter. 1. A lens prism or partial mirror that separates the reflected image light into RGB components. **2.** A lens prism system that reflects a portion of a light beam and permits the rest to pass through in order to separate colors or to produce two images in two separate locations.

bear trap. A sturdy set-spring clamp, usually a light mount. Also called a *gaffer grip*.

beat. 1. Accented notes in a musical composition used to provide desired rhythm. **2.** A directional word inserted in a film script dialog to indicate a momentary pause in an actor's speech (or occasionally action). Used to create a break in a lengthy monolog.

beater movement. A method by which a film is forced through a projector gate by a roller that beats down on the lower loop, as opposed to the claw and sprocket method.

Beaulieu. A small 8/16 mm French camera.

beauty shot. Close-up of a product being advertised.

beep. A brief sound track tone aligned with a visual reference point to help provide meticulous synchronization in editing and printing the film.

beeper. 1. A device used to introduce a brief tone or series of tones into an audio system. **2.** A sound used to inform a telephone caller to a talk show that he is being recorded.

behind-the-lens filter. A gelatinous filter in a metal container that can be placed in a camera slot between the film and the back of the lens.

bellows. A device used to extend the focal length of the lens. It is made of a section of lightproof pleated cloth formed in a square that is attached to the camera at one end and to the lens at the other end.

belly board. A flat board on which a camera can be mounted for low-angle shots.

below the line. Concerning program costs, a term referring to costs for technical rather than creative services.

bending. Visual distortion created by faulty videotape/playback head-timing coordination.

best boy. The chief assistant to the head GAFFER on the set.

bias. A high-frequency AC current applied to a magnetic recording circuit to reduce noise and distortion.

bias light. A characteristic of a lead oxide television camera pickup tube that wipes (or erases) an image in order to reduce HALATION or lag.

bicycling. The exchange between stations of film prints and videotapes, meaning the actual conveyance of the items from one station to the other.

bidirectional. Referring to a microphone that has a pickup response or pattern of front and back only. Can also refer to the pattern itself.

bifilar. An AC transfer videotape duplication that uses both master and single-dubbing record recorders.

big eye. A 100,000-watt floodlight.

big head. Slang for CLOSE-UP SHOT of actor's head.

bilateral variable-area sound track. A SOUND TRACK in which the modulations are symmetrical along the long center line of the track.

billboard. A sponsor message or identification shown near the opening or closing of a broadcast program.

billing. 1. A charge made to an agency or client for advertising time on a specific program or programs. **2.** The position of screen credits given members of a cast according to the terms of their contracts.

bin. A container holding unspliced film lengths in a cloth bag hung from a PIN RACK.

binary opposition. The juxtaposition of a pair of related opposites for dramatic effect. (E.g., the interplay between nature and culture represented by the protagonists in *The African Queen.*)

binaural. Referring to the placement of two distinct and separate audio sources in a recording, each heard by a different ear.

binder. A gelatin coating that holds light-sensitive silver particles to the tape base.

Bioscope. A type of motion-picture projector used in the pioneer days of the film industry.

bipack. 1. Two sections of film that are run together in order to be printed as one (double exposure). **2.** Camera used for this process.

bipack double-print titles. Film titles made by printing a duplicate from a bipack that shows both action and title films with the title superimposed over the action in either black or white depending upon whether the film is negative or positive.

bipost lamp, socket. A device that includes a lamp, with two grooved shafts or rods that extend from the bottom, and a socket that has two holes into which the lamp post is inserted.

bird. A satellite. Occasionally used as a verb, meaning to transmit via satellite.

bird's nest. The pile-up of film in a camera.

birthmark. See ADDRESS CODE.

bit. 1. A small role for an actor. **2.** Brief business used by an actor to enhance a scene.

bitch box. A small low-fidelity loudspeaker, used during audio recording, the purpose of which is to ascertain the response from a sample of home receivers.

bit player. An actor who plays minor roles in films.

black body. A substance that radiates and absorbs light with total efficiency.

black clipping. A video control circuit in cameras and videotape recorders that controls and contains the black level of the video signal so that it does not disturb (or appear in) the SYNC portion of the signal.

black comedy. A film that treats with humor serious subjects such as death, violence and racial prejudice.

black leader. Opaque unpunctured film used in preparing the original film for A and B roll printing with the printer light blocked.

black level. The darkest portion of a film.

Black Maria. The first motion-picture studio, built by Thomas Edison in West Orange, New Jersey; it was made to revolve in order to follow the sun.

black net. A black screen used to reduce light with minimum diffusion.

blackout. 1. The ban on live showing to home audiences of a local event (usually a sports event). **2.** A situation in which lights suddenly go out, by design or accident.

blackout switch. A master switch that controls all lights on a television set or sound stage.

Black Rock. An in-business term for the black granite headquarters of CBS in New York.

blacksploitation film. Any major film in which black actors are the principal, sometimes the only, performers.

Black Tower. The nickname for the administration building at Universal Studios in Universal City, California.

black week. One of four weeks annually in which Nielsen does not take rating responses from network television audiences. Sometimes called *dark week.*

blank. A transparent animation sheet, usually eleven by fourteen inches, used to maintain consistent density for filming.

blanketing. The widespread broadcast of a signal, exceeding 1 V/M (volts per meter), probably near the antenna.

blanking interval. The interval (10.5 microseconds) in which a television receiver scanning beam is eliminated by an electronic signal when returning to the left side of the screen to retrace the next horizontal path or to the top of the screen (1.3 milliseconds) to start another half of a scanning cycle (or field).

blanking level. The level that divides picture information from synchronization in a composite signal.

blasting. The sound of an instrumental or vocal performance that rises above the desired audio level.

bleach, bleach bath, bleach solution. A chemical liquid that removes the metallic silver from film EMULSION in the REVERSAL procedure, leaving intact the undeveloped silver salts that remain in the emulsion.

bleachers. Audience seats that can be moved in or out of a studio, to accommodate crowds of varying sizes.

bleed. To remove a portion of a picture.

bleep, blip. 1. A cueing tone signal on a sound track. **2.** To erase offensive or unwanted words on a sound track.

blend-line. The line that serves to separate a MATTE visual from an action visual when combined on the screen.

blimp. A camera case that is soundproofed to prevent noise by the camera motor from reaching the microphone.

blimping. The material used to soundproof the BLIMP.

blind bidding. The practice of selling exhibition rights to motion-picture theater owners before giving them a chance to see the films. (As of 1981, this practice had been forbidden in twenty-two states by the state legislatures.)

blind booking. Theater booking of a motion picture before it has been completed or shown to the exhibitors.

block. To determine the position of the camera, crew, and cast, and cast movements before shooting a particular scene.

block booking. A practice, no longer in use, of exhibitors being required to schedule inferior films in order to receive the desired feature films.

blockbuster. 1. In motion-picture parlance, a heavily financed major film, usually shown in a limited number of theaters at advanced admission prices. **2.** In television, a major network production given an unusual amount of advance advertisement.

block programming. A practice of television networks to schedule programs so that the audience will want to continue viewing from one to the other without switching channels.

blood. Artificial blood used in films, usually made from coffee, food coloring, and syrup.

blood capsules. Tiny capsules containing artificial blood that are stuck on small explosive caps, which, when exploded on or beneath an actor's clothing, make him appear to bleed from his "wounds."

bloom The dark area on a picture that creates a halo (or *halation*) around an unusually bright area.

bloop. 1. The sound that occurs in an amplifier and speaker system when a splice moves over the photo cell scanning slit connected to the amplifier. **2.** An opaque section placed over a splice to reduce or smother any undesirable noise on a positive film sound track. **3.** To delete any extraneous noise from a magnetic sound track with a small hand-held magnet. Definitions 2 and 3 are also referred to as *deblooping*.

blooper. 1. An oral error made by an actor or newscaster on a sound track or on live television. **2.** A punch or magnet used for deblooping. (See BLOOP, def. 3.)

blooping ink. Ink used to make a triangular-shaped hole over a negative sound track splice.

blooping tape. Tape used to cover sections of a sound track that are expendable.

blow. An actor's blunder in dialog or in action shots.

blower brush. A small brush used to clean the gate of the camera or projector.

blow up. 1. To make an enlargement from a 16mm to a 35mm film. **2.** To emphasize a detail on the screen.

blowup. 1. Any film print enlarged. **2.** Part of a frame enlarged to full-frame size in order to eliminate a portion or portions of the whole.

blue. Vulgar or obscene, referring to dialog or action.

Blue Book. The FCC regulations concerning Public Service Responsibility of Broadcast Licensees, published in 1946.

blue cometing. Pale blue spots in a developed color emulsion created in the processing bath by metallic contamination.

blue movie. Pornographic motion picture.

blue pages. Pages of film script that have been revised and inserted in the working script. (These pages may be any color but usually are blue.)

blue pencil. 1. To edit a film script. **2.** To censor a script.

blue-screen process. Action photographed in front of a blue background to create a MATTE effect.

blurb. 1. Short highly laudatory advertising material. **2.** A news release.

BMI (Broadcast Music, Inc.) A trade association of musical performers.

board. A panel or console operated by technicians in the control room.

board fade. Reduction of sound from the control room.

body frame, body brace, body pod. Braces on a camera that fasten to the shoulder and waist of the cameraman, releasing his hands for action.

body wash. Dark makeup used by actors.

Bolex. A Swiss 8/16mm camera operated by springs.

bomb. A motion picture or television program or series that fails to capture an audience.

bond. A safety cord or chain that holds a suspended LUMINAIRE.

boob tube. Pejorative term for television set.

book. 1. To employ performers or performances. **2.** The scheduling of a film by an exhibitor for a specific date and for an agreed-upon running period.

booking contract. The written agreement between an exhibitor and renter (or between actor and employer) specifying and legalizing details and conditions of the agreement.

boom. 1. Cantilevered mounts of various sizes and lengths on which film cameras are mounted. **2.** Lightweight metal hydraulic device used to suspend a microphone over filmed action and to move it from one location to another.

boomerang. A holder set in front of a LUMINAIRE that contains translucent filters that alter the color properties of light. (This filter is often referred to as GEL or gels.)

boom down or up. To change the position of the camera.

boom man. A film technician who operates a microphone BOOM.

boom shot. A shot made from a camera mounted on a boom for high angles or wide angles; a versatile shot that can be used to pan, tilt, or travel in all directions.

boomy, tubby. Sound with reduced definition due to resonance in the low frequencies.

booster. Technical devices used to enlarge and retransmit signals.

booster light. 1. Any artificial light used to augment daylight in outdoor on-location filming. **2.** To block out shadows or to strengthen details.

boot. A tube made of leather or fabric, open at one end to encase a tripod head.

booth. A soundproof studio on stage or set used for sound equipment and operator.

border light. A strip of lights containing five 1,000-watt bulbs.

Bosch. A German electronics company that produces the Fernseh handheld camera.

bottle. A glass container that holds the picture tube.

bottom pegs. The pegs on an ANIMATION BOARD that are closest to the crane operator as he faces it.

bounce lighting. Artificial light that is not aimed directly at the subject but is reflected on the subject by being aimed at the walls or ceiling, thus giving a diffused illumination.

box. Any set with four walls.

box office. 1. Small booth at theaters from which tickets are sold. **2.** Revenue from motion pictures.

box-office draw. A film or star that is successful in attracting audiences to the motion-picture theater.

B picture. A comparatively inexpensive film, usually intended to be a second feature on a double bill.

braceweight. A metal support for the BRACE.

bracketing. Exposing several sections of film at the light meter f-stop designation to ascertain the best exposure.

break. A brief respite, such as a coffee break, for performers. In television, time-out during a production for a commercial or station identification.

breakaway. A prop constructed flimsily so that it can be broken easily without harm to performers during action shots. (E.g., a breakaway chair or stair railing.)

breakdown. 1. Directions concerning action and dialog to be filmed in a specific shot or scene. **2.** The analysis of a film script for the purpose of determining all expenses for budget proposals. **3.** Rearrangement of shots or scenes to provide the desired continuity.

breakup. The brief distortion of a picture on the screen.

breathing. The loss of correct focus in an image as the picture seems to move in and out, due to the buckling of the film in the camera or in the GATE of the projector.

bridge. 1. A brief piece of sound track music to connect one scene or shot to another in a film. **2.** A sound or music that forms a connection between sections of a broadcast. **3.** A walkway that runs above a GRID.

bridge plate. A sliding plate on a camera mount used to correctly balance the camera.

bridging shot. A shot used for smooth transition between two shots or scenes when there is a break in time or other continuity.

brightness range. Variations in reflected light from objects or actors as indicated by a light meter.

bring up. To raise audio levels.

broad, broadside. A square wide-beam 2,000-watt floodlight that creates general and diffuse lighting on the set.

broadcast. The television signal transmission received by a wide audience of TV set owners.

Broadcast Bureau. A division of the FCC that serves in an advisory capacity.

broadcast homes. Homes in which there are one or more television sets.

Broadcast Pioneers. An organization of men and women who have worked for twenty or more years in radio or television.

B roll. See A AND B ROLLS.

Brute. The trade name for a large 10,000-watt spotlight used for illuminating locations that are unusually dark.

BTA (best time available). Used in television to indicate that commercials will be scheduled on the date or at the time chosen by the station.

buckle switch. A device used to stop the camera or projector when a loss of LOOP occurs during filming or projection. Also called *buckle trip*.

buckling. The curling of film that has been exposed to too much strain or heat, resulting in shrinkage.

budget. Funds required for film and television productions, derived at by listing all projected expenses for equipment, salaries, travel expenses, and other production costs.

buff. 1. Motion-picture fan. **2.** To remove scratches from film by polishing it.

bug eye. A wide-angle lens on a camera, designed for close-up shots. Also called *fisheye.*

build a mix. The consecutive addition of various audio and visual components to make a total amalgam.

built-in light meter. A meter that is a part of a camera and can automatically open and close the iris diaphragm to admit more or less light.

build up. To promote tension in a film by arranging shots that will build to a crisis.

bulb. 1. A glass bulb in which the TV picture tube is encased. **2.** A glass or quartz bulb in which lamp filament or fluorescent material is contained.

bulk erase. To erase a magnetic tape or film by realigning all iron dioxide molecules.

bullet hit. A small explosive charge placed behind or beneath a surface and electrically discharged to give the appearance of a bullet having struck and exploded on or in the surface.

bulletin. The announcement of an immediate news event, usually of a sensational nature, that interrupts regular broadcasting.

bull line. A thick rope used for manipulating scenery.

bumper. Referring to time given to persons using a studio in which to remove their equipment before another group arrives.

bump-up. A duplicate of a filmstrip, enlarged.

burn. An unwanted image held by the pickup tube target after it should have been replaced by another subject.

burn in. To superimpose a title over the projected action.

burned-out. Referring to any electrical equipment that has ceased to function.

bus. 1. A central interconnected electrical system. **2.** Manually manipulated buttons on a control room console.

business. Action used by actors to add dimension to their characterization or to a scene or shot.

bust shot. Film of an actor shot from the waist up.

busy. 1. Having in a scene an inordinate amount of action that is distracting. **2.** Having a fussy or ornate background that detracts from the foreground action.

butterfly. SCRIM that is tightened on an oval frame and used to diffuse strong sunlight on location shooting or to reduce any excessive light on a subject.

butt splice. A film splice in which there is no overlapping, the ends being joined neatly together by tape.

buy. Expression meaning to approve a performance or a performer.

buyout. Payment to an actor or musician on a single-performance basis, without RESIDUALS.

buzz track. A test film sound track used to correctly position the visual film in an OPTICAL SOUND reproduction system.

B-wind. Indicates the emulsion and perforation positions on rolls of film with single perforations. The flat side, vertically placed, is toward the operator and the film winds off the right side with the emulsion and perforation on the far edge.

BW print. The black and white print of a film, presenting the maximum contrast desired.

C

cable. 1. Flexible electrical wires in an insulated sheath used for transmission of electric power or signals. **2.** Heavy steel wire used in set rigging. **3.** Loosely, CABLE TELEVISION.

cable guard. Protective covering on the bottom of a DOLLY.

cableless sync. See CORDLESS SYNC.

cable news. Basically a twenty-four-hour news service offered to television homes that subscribe to the service. See CABLE TELEVISION.

Cable News Network (CNN). An all-news television cable network.

cableporn. X-rated movies offered nightly by as many as 200 cable systems to home subscribers.

cable ramps. Protective wedges that prevent cables from being damaged by the traffic on the set or location.

cable release. Flexible sheath encasing a shaft used to start the running mechanism on certain motion-picture cameras.

cable television. 1. A master antenna system with a head-in at a central location from which signals are picked up and transmitted along lines to individual television sets for which special equipment is installed and for which a monthly fee is paid by subscribers. Used to improve reception in places where transmission is not clear (often in low country surrounded by high terrain); used to provide satellite service from distant parts of the continental United States; used to provide premium (pay) television service to subscribers who have purchased the particular service. **2.** A service owned by a network that transmits programs of more specific audience interest (cultural and informational) than in general non-cable network programming, usually supported by advertising; sometimes owned and operated by towns and cities.

cadmium sulphide meter. A special light meter that uses cadmium sulphide as the light-sensitive factor.

calculator. A plastic or cardboard instrument used as an aid in figuring data concerning film running time, the correct infiltration for lighting of color temperatures, and the F-STOP compensation for filters.

calibration. Marking to indicate a specific measured point or the manner in which such points or positions are computed.

calibrations. Position marks on animation art backgrounds denoting the amount of movement between frames.

call, call board, call sheet, shooting call. A timetable, either published or posted, in which actors are instructed when to appear for a performance or rehearsal.

Callier effect. The dissipation of light rays as they pass through various components.

call letters. The alphabetical letters used to identify television stations.

cam. A rotating projection on a wheel within a film mechanism, which adjusts the forward flow of the film.

cameo lighting. Set lighting of actors in the foreground, photographed against a dark background.

cameo performance, cameo role. A minor acting part but one of importance to the film, usually played by a well-known actor.

cameo staging. Foreground action photographed against a simple unobtrusive background.

camera. 1. The apparatus, optical or electronic, consisting principally of a lens attached to a box that excludes all light, a mechanism to advance the film, a shutter, and viewfinder, used for the exposure of film. Some motion-picture cameras are also equipped to record sound. **2.** A transmitting apparatus in which the picture to be televised is formed before it is changed to electric impulses for broadcasting.

camera angle. Simply, the relationship between the camera and the subject(s) of the shot; it may be tilted up toward the subject for a low-angle shot, down from the subject for a high-angle shot. Without tilt, the shot is seen at eye level. Camera angles can also be interpretive; small subjects can be made tall, and vice versa; rooms can be enlarged or decreased in size, etc.

camera body. The stationary part of a camera that houses the drive mechanism exclusive of the lenses, motors, and magazines.

camera car. A car or truck especially equipped to carry the camera and cameraman for shooting while the vehicle is in motion.

camera card. A card that carries program titles or credits.

camera department. The studio personnel or division responsible for the storage and maintenance of camera and camera equipment.

camera identification mark. A manufacturer's mark placed next to the APERTURE, an opening in the plate that permits the area of film just behind the mark to be exposed simultaneously with each exposed frame. The exposed film can then be identified as coming from a camera of a specific make or model.

camera jack. A connector device for power or sound cables that can be attached to a camera.

camera leading. A term applied to a moving shot in which the subject actually follows the camera as it TRACKS backward.

camera light. A light mounted on a camera to provide close illumination of a performer.

cameraman. The principal camera technician who is responsible for the visual results of the shot, in both lighting and photography.

camera mount. A support for a camera that allows it mobility for a PAN SHOT or TILT SHOT; can be attached to a DOLLY, BOOM, or TRIPOD.

camera movement. The motion of the camera used to achieve special cinematic effects, such as the TILT SHOT, PAN SHOT, CRANING, DOLLY SHOT, TRACKING, or ZOOM SHOT.

camera noise. The whirring sound of a camera, an unwelcome noise during sound recording.

camera obscura. A room or box with a lens set into one side, through which outside objects can be seen on the inside wall opposite the lens.

camera rehearsal. An actor's rehearsal during which camera movements are blocked for the forthcoming scene or shot.

camera report. The report made by the camera crew that records shooting data, take-by-take, and the date, names of the crew members, film emulsion number, filters, camera number, footages, and notes on good and bad TAKE numbers; enables editors and laboratories to determine which shots are to be printed.

camera right, camera left. Directions needed by directors and cameramen in blocking the scenes into desirable patterns of motion.

camera speed. The rate at which film runs through a camera, measured by frames per second or in feet or meters per minute.

camera stylo, camera pen. A term originated by French director Alexandre Astrue, describing his belief that a motion-picture camera can be used by a cameraman in the same way a pen is used by a writer, as a means of artistic expression.

camera talk. Term used to describe a shot of a person who stands directly before the camera, looking into the lens, usually to introduce a film or deliver a public-service editorial or a political message.

camera test. 1. A test made of a camera's performance, to check on speed and steadiness. **2.** A short negative end section of film exposed for testing in a laboratory.

camera trap. A place in which a camera is concealed during filming.

camera tube. A pickup tube that converts optical images into electrical signals through an electronic scanning operation.

cam intermittent. A kind of film-flow mechanism that uses a CAM and CLAW.

campaign. The advertising schedule for a specific product during a specified time period.

can. A metal or plastic container used for the transportation and/or storage of film.

candela. The measurement of the intensity of light discharged by one-sixtieth of a square centimeter of platinum at a temperature of 2042 degrees Kelvin.

canned. Prerecorded.

canned laughter. Recorded laughter played to encourage audience response during comedy routines or situation comedies. Also called *sweetening.*

canned music. Recorded music; background music provided by a recording.

Cannes Film Festival. An international film festival held annually in Cannes, France, in which hundreds of films are screened, both in and out of competition.

cans. Headphones.

canting. Angling a camera position.

cap. Cover for a camera lens.

capacitance. The property of an electric nonconducter that permits the storage of electric energy in an electric field, measured in farads.

caper film. A motion picture in which the plot revolves around a major theft or other illegal enterprise, usually a highly imaginative and complicated maneuver; developed from the point of view of the lawbreakers.

capping shutter. A camera shutter that can be closed without use of the exposure shutter, used on certain animation cameras to block out undesirable light.

capstan. A motorized shaft, usually operated with a pinch roller, used to transport magnetic tape through a recorder.

capstan servo. A helical videotape recorder head phase and tape speed control system that provides correct sequential readings of video information.

caption. Words superimposed over action in a shot, providing necessary information to the audience.

captions. Dialog superimposed at the extreme bottom of the frame, usually as subtitles for foreign-language films. Also provided for hearing-impaired television viewers or occasionally when spoken words are so garbled as to be unintelligible.

cap up. To cover the camera lens.

carbons. DC arc lights.

cardioid microphone. A microphone with a heart-shaped unidirectional pickup sensitivity area.

card rate. Broadcast station's fees for commercials, according to the time the commercial is given, the length of the message, and how often it is to be used.

carnet, tempex. A customs form needed in order to transport film equipment for temporary use on locations in Europe.

carrier. The frequency wave that transmits television signals.

CARS (Community Antenna Relay Station). A specific microwave frequency (12.75–12.95 MEGAHERTZ) band that the FCC has assigned to the cable television industry for use in transporting signals to cable system HEAD ENDS.

cartoon. An animated short subject made from drawings or paintings.

cartridge. Any container holding film or tape that can be inserted easily without threading, usually applied to factory-packaged containers that require no extra installation by the consumer.

cartridge camera. Any camera that uses cartridges of film rather than film wound on spools or cores.

cassette. A small tape cartridge containing two reels of one-eighth-inch magnetic tape, one to feed and rewind, the other to reel up the tape.

cassette recorder. A magnetic sound recorder in which cassettes of packaged tapes are used.

cast. 1. The performers in a film. **2.** A title list of the actors, which is shown among the credits preceding or following a motion picture. **3.** To hire performers for a film production.

cast commercial. A broadcast advertisement of performers for a specific program or series in which they appear.

casting director, cast director. The person who selects the performers for a film, usually with the exception of the principal actors.

catadiatropic lens. A telephoto lens with mirrors designed to reduce the size of the lens.

catchlight. 1. Tiny specks of light appearing in the eyes of a person being filmed. **2.** Small lights on or near the camera to obtain that effect.

cathode. A negatively charged TERMINAL.

cathode ray tube. A device that contains an electron gun that discharges controlled beams of electrons against an internal phosphorescent screen. A television picture tube is a CRT.

cattle call. A general talent audition from which minor roles, usually for singers and dancers, will be filled.

CATV. See COMMUNITY ANTENNA TELEVISION.

catwalk. A narrow walkway suspended over a studio set on which electricians and stagehands can walk in order to hang or service lights and sound equipment.

CBC (Canadian Broadcasting Corporation). The state-controlled network of Canada.

CBS (Columbia Broadcasting System, Inc.). One of the three major American television networks.

CCR. Central control room in a television studio.

CCTR. Closed-circuit television. A system for the transmission of television signals that are not broadcast over public channels.

CDL. A computerized editing system for videotape.

cel. A transparent rectangular plastic sheet, usually cellulose acetate, which supports drawings, either inked or painted, such as those in titles and animation work. Contains punched holes into which pegs are fitted in order to facilitate the registration of successive cels during photography.

cel animation. A motion-picture film made from drawings or graphics produced on CELS.

cel flash. A hot spot (unwanted reflection) caused by a cel surface that is not uniform.

cel sandwich. Two or more CELS placed together to provide a necessary combination of graphics.

celluloid. **1.** A flammable film base made of cellulose nitrate. **2.** Slang expression for a motion picture.

cellulose acetate, cellulose triacetate. The transparent flexible film used as the base support for photographic EMULSIONS, as well as in sheets used in animation and title work.

cement. A liquid solvent used to splice two ends of severed filmstrip.

cement splicer. A device used to overlap two ends of film and hold them securely as the cement is applied to seal the overlap.

censor. A person or group that views films before they are shown to the general public in order to determine their merits, particularly in regard to objectionable material.

center of perspective. The visual position from which the image in a film shot will coincide dimensionally with the actual subject.

center track. A standard position for the audio signal on double-perforation magnetic film, placed in a narrow band halfway between the edges.

centisecond. One-hundredth of a second.

century stand. A LUMINAIRE metal support with three telescoping legs.

changeover. The transition from one reel of film on a projector to the next reel on a second projector to achieve an uninterrupted flow of film during a feature-length film presentation.

changeover cue. A circular mark on the upper right-hand corner of several frames, displayed during the final frames on a motion-picture reel to warn the projectionist that it is time to switch from one projector to another.

changing bag. A light-proof cloth bag with armholes that allows the operator to load film outside a darkroom without fogging.

channel. **1.** A signal circuit devised for pickup, transmission, and/or electrical control of sound. **2.** FCC-assigned waveband frequency (in the United States the spectrum space is about 6 MHz wide for each channel). **3.** The output of an individual television station.

character. A role in a film.

character actor. An actor who can play a variety of roles, rarely the principal ones, and usually roles of older persons.

characteristic curve, D log, E curve, contrast curve. The curve that occurs when the densities in a developed photographic EMULSION are measured in areas where the exposure was gradually increased and plotted at coordinate points on graph paper.

characterization. An actor's projection of a role that concentrates on the personality, background, and character of the person portrayed. Usually results from the combined efforts of the screenwriter, the director, and the actor.

charger. An electrical device that replenishes power in batteries.

chase. Any pursuit in a motion picture that creates suspense, usually dependent upon rapid and dangerous action such as a car chase through crowded city streets.

chase film, chaser. **1.** Any film in which a principal plot device is a sustained chase (e.g., *Les Misérables*). **2.** A sequentially wired row of lamps that provides the illusion of movement in light. **3.** The music played during a performer's exit.

cheat. **1.** To arrange in unrealistic positions a group of PROPS or subjects before a camera so that they will enhance the composition of the shot. **2.** To use an APPLE BOX or HALF-APPLE to create an illusion of height: e.g., "Cheat the shot. Put the short man on a half-apple."

cheat shot. A shot in which the action is not as it appears to be. (E.g., illusion that an actor jumps from a cliff and lands in a raging river, whereas he actually lands in a net placed just beneath the cliff.)

checkerboard cutting, checkerboarding. **1.** Film-editing technique used in splicing AB ROLLS so that the film splice image will be eliminated from the duplicate. **2.** Scheduling programs every other day or every other week.

check print. An unedited film print made from a negative and used to check any mechanical printing errors.

chemical fade. A kind of fade made by immersing film in a chemical dye so that the scene is darkened until the image is fogged over.

chemistry. **1.** The combination of temperatures, time elements, and sequences used for the development of film. **2.** Photo development chemicals.

chest shot. A shot of a performer from the waist up.

chicken coop. A LUMINAIRE that is surrounded by a protective wire-mesh screen.

chief engineer. The technician who is in charge of the control room.

children's film. Any feature film designed for the entertainment of children.

china girl. The negative film image of a white American girl's face, used as a color standard by all United States film laboratories.

chinese. Referring to a photography technique in which the camera pulls back and PANS simultaneously.

chip. A filament discharged by a cutting stylus.

chip chart. A test-swatch chart in black and white used to check the alignment of the camera.

choreographer. A person who creates and directs the movements in a dance, such as a ballet or musical comedy routine.

Christmas tree. Several LUMINAIRES mounted on a single stand.

chroma. The measure of the saturation of a color.

chroma control. The color control on a television receiver.

chromakey. An electronic matting technique in which the subject matted is placed against a solid-color background (usually blue) and the signal mixed with that particular color channel is suppressed.

chromatic aberration. A lens defect that creates different focal points as color fringe halos.

chrominence. Color camera channels for red, green, and blue (RGB) signals.

chromium dioxide. A noncompatible audiotape coating with an improved signal-to-noise ratio.

churn. The turnover among cable television subscribers.

CHUT (Cable Households Using TV). An audience survey estimate of the separate households viewing television during an average quarter-hour period.

cinching. Tightening the film on the end of a reel.

cinch marks. Vertical scratches on film made by dust or other particles between coils, or by tight negative winding.

Cinderella film. A modest-budget film that has an unexpected and overwhelming success at the box office.

ciné-. Prefix designating motion pictures or film.

cinéaste. A person who studies films or the making of films, more serious and dedicated than the average film buff.

ciné board. 16mm footage of STORYBOARD FRAMES, edited against a sound track. (TV)

ciné camera. Any camera used in making motion pictures.

ciné 8 film. 8mm motion-picture film with only one perforation on one side at each frame line.

ciné-fi sound. Sound accompanying a motion picture at specific drive-in theaters, heard by dialing an indicated AM station number on the car radio; if the radio cannot be played unless the ignition is on, a portable AM radio can be used.

cinema. A motion picture, motion pictures collectively, or a motion-picture theater. Frequently used as an adjective, such as "cinema critic" or "cinema buff."

CinemaScope, Scope. The Metro-Goldwyn-Mayer trademark name for a wide-screen film process using ANAMORPHIC LENSES on both camera and projector.

cinematic. Usually referring to the various factors in filmmaking that are not included in the actual filming process: e.g., the distribution, financing, and budgeting of a film are parts of the cinematic process.

cinematic synthesis. A basic principle of film animation in which film is shot frame-by-frame, with each frame projected as part of continuous motion, as in STOP-ACTION PHOTOGRAPHY; used for SPECIAL-EFFECTS sequences.

cinematographer. The director of motion-picture photography who is responsible not only for the photography but also for the lighting and the technical elements involved in setting up the shots.

cinematography. Motion-picture photography.

cinéma vérité. (From the French, meaning *cinema truth.*) A filmmaking genre started in Europe in the 1950s. Portable sound cameras were used to record events and interviews on location sites, often with persons who were not actors, and presented in a documentary style for films in which radical political views were often expressed.

cinemicrography. The shooting of a motion-picture action through a microscope attached to the camera.

Cineplex. A complex of several motion-picture theaters that offer first-run, foreign, art, and specialty films simultaneously.

cineradiography. The process of producing x-ray films with a variety of techniques, usually involving image intensifiers.

Cinerama. A wide-screen film process that originally employed three cameras, three projectors, and stereophonic sound to show the image in overlapping panels on a large curved screen. The later form uses a single projector with 70mm film.

cinesemiotics. The study of motion pictures as a system of signs with denotative, connotative, and contextual meanings.

Cinex printer, Cinex machine. A device that prints laboratory test strips with graded exposures to determine the correct final RELEASE PRINT color selections for the shot.

Circarama. A 360-degree screen film process, in which the audience is entirely surrounded by the film.

circled takes. The shots to be WORKPRINTED, indicated by the circled numbers on a CAMERA REPORT.

circle of best definition. A circled portion in the CIRCLE OF ILLUMINATION that contains image definition that is approved for photography.

circle of confusion. The circular image of an object point on a section of film; the size depends on the correct lens focus. In 16mm photography when the circle of confusion exceeds one-thousandth of an inch the subject will become smudged.

circle of illumination. The portion used for photography of the round image formed by a lens on a surface behind the lens.
circuit. 1. Electrical system. **2.** Chain of motion-picture theaters owned by an organization.
circuit diagram. A chart used to diagram the connections among components and parts in an electrical or electronic device.
circular polarization. A right-handed corkscrew television signal transmission pattern in which reflected secondary picture tube images are minimized by a left-handed polarity shift.
clam bake. A program that has been produced with less than professional skills.
clamping. Determining a fixed-reference DC video signal at the start of each SCANNING LINE.
clamping disk, knuckle. An adjustable head for a CENTURY STAND, grooved to accommodate FLAG STEMS, PIPE BOOMS, and other equipment.
clapboard, clapper board, clapstick board. A slate board on which information relative to the film (such as director, cameraman, interior/exterior, shot number, and take number) are written. It is photographed at the beginning of each shot. See CLAPSTICKS.
clapper. See CLAPBOARD.
clapper boy. The technician who operates the CLAPBOARD.
clapsticks. Two boards hinged together at one end and connected to the top or bottom of the CLAPBOARD. They are slapped together to provide an audible cue in the sound track. With the clapboard, they provide cues for visual and sound synchronization in editing.
Clarke process. The filming of action through a dispositive plate, a portion of which is clear and a portion of which contains a still image.
class A, B, C, D. Designations of time periods for broadcast commercials, according to the size of the audience.
classification. A system used to determine a film's suitability for audiences of various ages. The classifications, provided by the Motion Picture Association of America, are: G—general audience; PG—parental guidance advised; R—restricted for persons under seventeen years old unless accompanied by a parent; and X—persons under seventeen years old not admitted.
claw. A camera/projector mechanism that intermittently advances single frames of film into a film GATE while the shutter is closed.
clean entrance. The movement of an actor from out of camera range into the action area during a shot.
clean exit. The movement of an actor from the action area out of camera range before the end of a shot.
clear. 1. To arrange the transmission of a network program with its affiliates. **2.** A filter used as a protective lens covering. **3.** To chemically remove bleach or undeveloped silver from film in the processing procedure.

clearance. Permission to use copyrighted material in a film.

clearing bath. A chemical wash used to remove bleach or undeveloped silver from film being processed.

clear the frame. A request to vacate the area in front of the camera during rehearsal.

click stops. A camera device used to stop and/or hold a movable part at a calibrated position. See CALIBRATION.

click track. A SOUND TRACK on which a series of clicks has been recorded, picked up only by the earphones of a musical group's conductor. Used to achieve correct tempo for the post-recording of the sound track that will accompany a film.

client. 1. An agency or advertiser who buys broadcast time. **2.** A person or organization for whom a film production company performs services.

cliff-hanger. A suspenseful action film, with suspense intensified at the climax. Name comes from the early serials in which actors were left in precarious situations at the end of each segment, to be rescued in the next one the following week.

climax. A moment of high tension, the dramatic peak at the end of the film, in which the conflict must be resolved.

Clio. A statuette for the best television commercials awarded annually by judges of the American Television and Radio Commercials Festival.

clip. 1. A short segment of a long film. **2.** A film insert used in television programming. **3.** The plastic or metal fasteners used to join two ends of film rolls without splicing. **4.** The inadvertent omission of a musical note or word at the beginning or end of a sound track.

clipping. 1. The removal of a signal portion above or below the pre-set level. **2.** An illicit operation engaged in by some stations of substituting local commercials for network commercials at double the usual payment.

clogging. Superfluous tape oxide on a recording or playback head, resulting in improper TRACKING or damage to the tape.

closed-caption, closed-captioned. A system that enables hearing-impaired viewers to "read" programs through captions that appear on a television screen by the use of a special adapter on the set.

closed-circuit. Referring to television programs that are not commercially broadcast but are distributed by cable to specific audiences.

closed set. A television studio set used for private filming or taping sessions.

close-up shot, close-up, close shot, CS. A film shot in which an object or actor is photographed so that it or he/she fills most of the frame. A close-up of an actor would usually include only his face and shoulders or a portion thereof.

cloud wheel. A device used to project a sky effect on a large white background.

cluster bar. A mount used to hold several LUMINAIRES.

clutter. The transmission of an inordinate number of television commercials and promotional material (sometimes as much as 25 percent of prime broadcast time).

C mount. A device used to mount a lens on a 16mm camera.

coated lens. Lens having a magnesium-fluoride coating on its air surface in order to reduce reflection by increasing the transmission of light rays to the film coating.

coating. The process by which emulsion and oxides are spread smoothly on film bases. Also, the emulsion or oxide after it has been applied to the base.

coaxial, coaxial cable. A three-quarter-inch television signal transmission that carries thirty to forty channels with low power loss at high frequencies; has repeater amplification every one-third mile.

coaxial magazine, concentric magazine. A camera magazine in which the supply and take-up rolls of film are opposite and parallel to each other.

cobalt-energized. Referring to a compatible audiotape coating with improved signal-to-noise ratio.

cobweb spinner. A device used to produce cobwebs; it consists of an electronically operated fan and a supply of rubber cement that is blown on walls, ceilings, and furniture to create the desired effect.

coded edge numbers, coding. Any system used to mark two or more films with the same series of numbers in sequence, in order to maintain a synchronous relationship.

code generator. Equipment that records visual identification signals onto videotape.

coercivity. The amount of magnetic energy, measured in oersteds, needed to produce normal videotape particle patterns.

cogwheel effect. A staggered vertical image produced by the microsecond displacement of alternate scan lines.

coincidental interview. A telephone audience survey in which persons are asked about the television programs they may be watching at that time.

cold. Without rehearsal; unprepared; referring to an actor's performance.

cold lights. Fluorescent lights.

color. **1.** Natural color photographed by a camera. **2.** Anything used in a film to provide a realistic atmosphere.

color balance. **1.** The relationship among color elements. **2.** The adjustment of the relative sensitivity to light of various colors to provide acceptable film or television pictures.

color-balanced. Referring to the process by which a film emulsion permits pure light of a designated color temperature to appear in the final print as white.

color bars. A bar-shaped videotape leader test pattern, electronically produced, which matches the playback to the original recording levels and phasing.

color-blind film. Film that is sensitive to only a portion of the color spectrum (as to blue but not to yellow). This quality permits the use of yellow SAFELIGHTS in film printing or processing rooms.

color burst. The 3.58 MHz subcarrier frequency primary color relationship sample at the back of each scan line which synchronizes transmitted color to a receiver. Timed to a quarter-millionth of a second.

color cast. An overall tint, such as pink, in a film image, usually undesirable but sometimes produced purposely to achieve a particular effect.

color-compensating filter. A lens filter used to achieve overall color balance; used on cameras or in PRINTERS to make corrections in the color of light being used.

color-conversion filter. A lens filter used to effect a large change in the color temperature of the light being used.

color correction. The readjustment of tonal values of colored objects or images by light filters, in order to meet the camera and lighting requirements.

color duplicate negative. A print with a negative color image made from the negative original, used for RELEASE PRINTS in order to protect the original.

colorer. See OPAQUER.

color film. Film with one or more EMULSIONS in which the light values of a scene are translated into various colors after processing.

color-film analyzer. An electro-optical device that scans a color negative in order to determine the exact exposure for a desired print.

color internegative. A negative-image color print made from a positive color print; used to make RELEASE PRINTS in order to protect the original. Contains visual effects made by AB ROLL PRINTING of the original.

colorimetry. Analysis of the technical qualities of electronic color-reproduction equipment.

colorizer. A device that mixes color signals with LUMINAIRE signals, resulting in bizarre color effects.

color master positive. A positive color print made from a negative color original, used to make 35mm color duplicate negatives and 16mm color and/or black and white duplicate negatives.

color match. The exact reproduction of colors as they appear on film from shot to shot or from one reel to the next.

color media. Any transparent materials, such as glass, which are used in front of the source of light to alter colors.

color print. A positive-image copy of a color original.

color-print film. Film made especially for use in making positive prints from color originals and color intermediates.

color processing. The development of film by use of chemicals in order to produce color images.

color-reversal film. Film used to produce a positive color image when processed after camera exposure.

color-reversal intermediate (CRI). A color duplicate negative made by the

REVERSAL PROCESS from a color original.

color sensitivity. The degree to which a photosensitive EMULSION responds to exposure by the wave lengths in white light.

color separation. The way in which black and white color negatives are prepared, one for each primary color, from an original subject or a positive color film.

color shift. An undesirable change in color appearing on the screen.

color television. The effect of total color produced by the transmission of three separated primary color (red, blue, and green) signals that are superimposed at the receiver.

color temperature. The temperature to which a black body (a substance that radiates and absorbs light with total efficiency) must be raised from absolute zero to emit light of a certain color. Various colors require different degrees of temperature.

color-temperature meter. A meter that indicates the COLOR TEMPERATURE of a light source, expressed in degrees centigrade beginning at zero.

Color Tran. Manufacturer's name for an autotransformer system used to increase the voltage in standard lamps, thereby raising their COLOR TEMPERATURE.

color video analyzer. A device used to determine the colors present in a film shot and indicate the additive PRINTER LIGHT required to expose a duplicate.

combined move, compound move. Simultaneous movement of actors and cameras during a shot.

comeback. The return of an actor or filmmaker to prominence after an absence or decline in popularity.

comedy. A film in which the subject matter is intended to amuse the audience.

comedy of manners. A comedy that presents characters in a way that mocks their social behavior.

come in. Instruction given to cameraman to move the camera closer to the subject.

cometing. The appearance in a developed EMULSION of small light spots made by foreign metallic substances in the processing bath.

comet tail. A smear on a television screen caused by a moving HOT SPOT or light source.

coming up. A reference to the program that follows the one being viewed, or one to be shown in the near future.

commentary. **1.** Editorial remarks delivered by a station executive or remarks made by a broadcast news analysist. **2.** Narrative that accompanies a documentary film.

commentator. News analyst.

commercial. Paid broadcast advertising, usually designed to sell a product or a political candidate or issue.

commercial broadcasting. Programs paid for by sponsors with products to sell.

commercial program. Any program that contains advertising material.

commercial protection. See PRODUCT PROTECTION.

community antenna television. A service that transmits broadcast programs, original programs, and services to subscribers in a wide geographical area through coaxial cable connections from a master antenna.

compandor. A device that compresses and/or expands voice modulation.

compatibility. The ability of a black and white television set to receive color signals with only a small amount of distortion in the picture.

compilation film. A film made by editing and cutting an excessive amount of footage shot without a script.

completion services. Work on a motion picture that is done after the actual filming, such as processing, printing, title and credit designs, and additional sound.

component television. Home systems composed of separate screen, tuner, and speakers used for better overall performance.

composite. **1.** A film print that contains both audio and visual effects. **2.** A film print that combines the images of two or more MATTE rolls. **3.** A single STILL that combines photographs of an actor taken from different angles.

composite master. An original videotape with both audio and visual effects.

composite matte shot. A film shot with a composite picture achieved by double or multiple exposure in the camera, using desired matting for each exposure.

composite photography. A special-effects technique in which two images seem to appear together in the same shot despite the fact that they are photographed separately; originally called *trick photography*.

composition. **1.** The relationship between masses and degrees of light and color in a film shot. **2.** The manner in which the actors and other elements of a scene are arranged within the shot.

compound. The part of an ANIMATION STAND that facilitates the placement of artwork.

compression. The reduction of a signal within a specified amplitude at the output in order to increase a signal level at the input; this prevents distortion in optical recordings.

computer animation, computer-generated animation. Animation filmed by techniques that use computers and associated RASTERS to control the images.

computer graphics. The animated portion of a film (using live actors) that has been generated by a computer programmed to produce a variety of drawings with precision timing; used in fantasy films such as *Star Wars* and *The Black Hole*.

computer interpolation. A method of producing graphics which are used in special effects in fantasy films; one image is turned into another through computer programming.

concave lens. A single lens that has one or two concave spherical surfaces.

condenser. A spotlight-focusing lens.

condenser microphone. A microphone in which one plate of a capacitor acts as a diaphragm whose vibrations create variations in voltage that can be used with IMPEDANCE controls as an audio signal.

condensing optics. Camera lenses that are used to concentrate narrow light rays.

conductor. **1.** Medium transmitting electric current. **2.** The leader of an orchestra or other musical group.

cone, cone light. **1.** A vibrator component of certain loudspeakers. **2.** A floodlight that emits a widespread beam of diffused light, 750–5,000 watts.

confirmation. The formal acceptance of broadcast advertising by a station.

conflict. The struggle between opponents (hero vs. villain, man against nature, etc.) that is a necessary component of any dramatic plot.

conforming. **1.** The manner in which original film is matched to the edited WORKPRINT. **2.** The manner in which original film rolls are arranged into A AND B ROLLS.

conkout. The failure of equipment used for filming.

connector. A link composed of two matching parts used to join or separate electrical circuits.

console. A specialized computer terminal in a television station control room into which broadcast elements are fed and controlled by switches that operate both recorders and projectors.

console editing machine. See FLATBED EDITING MACHINE.

contact print. A positive film printed from a negative by direct emulsion-to-emulsion contact when run through the printer.

contact printer. A printer on which two pieces of film, the raw stock and the film to be printed, are run through with their emulsions in direct physical contact.

CONTAM (Committee on Nationwide Television Audience Measurement). A network group that oversees the audience-rating practices.

contamination. **1.** The pollution of photographic chemicals caused by foreign elements such as dust, other chemicals, or metallic substances. **2.** The faulty separation of various color signal paths.

continuity. **1.** The even flow of events in a dramatic film production. **2.** Script that contains all necessary visual and audio instructions in the final pre-shooting phase, subject to change only by the director. **3.** An even transition from shot to shot.

continuity clerk, continuity girl. The technician who keeps a record of the details of every take each day so that continuity can be maintained without errors when the shots appear in correct sequence.

continuity cutting. The manner in which a script is edited to maintain a smooth flow of events in logical sequences.

continuity sketches. An artist's drawings for a particular script, used to serve as a guide for the composition of shots.

continuous action. Action uninterrupted during the filming of a scene or several shots filmed by two or more cameras set at various angles. On the screen the action flows evenly from one camera to another or one angle to another.

continuous contact printer. A contact printer in which the film moves steadily.

continuous exposure. The use of a light beam that flows constantly in a printer without interruption by shutter action.

continuous optical printer. An optical printer in which the film moves steadily.

continuous printer. A printing machine that moves the film past the exposure aperture in an even flow and with no interference with its rate of advance.

contract. A written agreement between employer and employee (for example, studio and actor) or between management and union, with stated terms concerning salary, work schedules, and work conditions.

contract film. A film made under a formal written agreement for which costs are to be paid for by the client, such as a production company or studio.

contract negotiations. Bargaining between employee (production company) and the actors and production personnel to determine salaries, credits, RESIDUALS, and other matters involving expenses.

contractor. The director or supervisor of a musical group.

contract player. An actor who does not receive a regular salary but who is employed for a particular job and for a limited time.

contrapuntal sound. Sound that is in direct contrast to action seen on the screen (for example, laughter or disco music heard during a funeral service at a cemetery).

contrast. 1. *Luminance contrast* is the ratio between the lowest and highest intensities. **2.** *Photographic contrast* is the ratio between the highest and lowest densities. **3.** *Subject* or *object contrast* comes from the short and long tonal scales with gradations from white to black. **4.** In a script, contrast is the presentation of opposites (a manor house and a hovel, beauty and the beast) for heightened dramatic interest.

contrasty. A visual that lacks middle tones and has an unusual amount of dark and light areas.

control panel. A console from which studio lights and certain apparatus are controlled electrically.

control ring. A mechanical device used to raise or lower the pedestal of a television camera.

control room. The soundproofed room in a studio from which the production management overlooks the set.
control strip. A section of film containing a series of measured exposures in which the density is checked after development to determine the extent of the development and the plotting of a contrast curve.
control track. A portion of a videotape signal that controls synchronization during the reproduction of previously recorded material.
control wedge. See CONTROL STRIP.
convergence. A three-color beam crossover focused electronically at the aperture mask of a camera.
convergence pattern. A television test signal used to check the monitor picture against distortions.
convergent lens. A camera lens, such as a plano-convex or bi-convex lens, used to create positive images.
conversion filter. A lens attachment used to allow indoor film to be used outdoors, and vice versa.
converter. **1.** The equipment used to convert television signal characteristics for international use by translating the signals from one national standard to another. **2.** A device that permits a television receiver to receive additional channel transmission.
convex lens. A camera lens that has one or two convex or extruding sides.
cookie. A cut-out opaque sheet placed in front of a light source to produce shadows in the action area.
cool. A television picture that has a slight blue or green cast.
co-op. The cost of broadcast advertising material shared by the manufacturer and the local merchant or distributor.
copter mount. A device used to attach cameras to helicopters for aerial shots, designed for minimum vibration and accurate focusing.
copy. Script written for advertising material/commercials.
copyright. A legal document establishing the ownership of an artistic creation such as a literary work, musical composition, photograph, or motion picture; may be obtained by registering it with the United States Register of Copyrights; claim to copyright must appear near the title of the work.
cordless sync. Any system used to achieve synchronism, without a connecting SYNC-PULSE CABLE, between the camera and the tape recorder.
core. A plastic cylinder used for winding and storing film.
core-to-core. Winding a film from one plastic core to another.
coring out. The process of erasing signal noise by digital conversion.
corporate campaign. Broadcast advertising designed to self-promote a company itself rather than the products it manufactures or produces.
Corporation for Public Broadcasting (CPB). The organization that distributes federal funds to public radio and television stations, commissions some programming for both, and operates under the Public Broadcasting Act of 1967 to advance noncommercial programming.

corrective commercial. Broadcast advertising information used by advertisers to correct some erroneous or misleading information in a previous commercial; done at the orders of the Federal Trade Commission.

cosmetic makeup. Makeup applied to improve an actor's appearance without making major changes.

costume designer. The person who creates costumes worn by actors in order to create or maintain an atmosphere and reveal personality or character.

costume film. Usually a historic film in which the costumes are an integral part of the atmosphere of the period depicted.

costumer. The person in charge of acquiring and fitting the clothes worn in a film, by having them made, buying them ready-made, renting them, or obtaining them from the studio's wardrobe department.

costume house. An establishment that deals in costume rentals.

costumes. Clothes worn by performers in a film.

countdown. A cueing signal, from 10 to 2, shown on videotape or film leader in intervals of one second.

counter. A device with a dial or face on which lengths of film can be counted in frames, six-inch increments, or feet.

counter commercial. A broadcast advertising message that counters or corrects some misleading commercial information, usually presented by a public-interest organization or group.

counter-key. A source of illumination placed opposite the KEY LIGHT.

counter matte. The matte used in second exposures of an in-camera matte shot to prevent double exposure of the pre-exposed areas.

counter programming. Program scheduling arranged by a network to counteract the strength or drawing power of the programs scheduled by another network during the same time period on the same day.

coupler. A device used to unite two pieces of cable that contain the same electrical elements.

cove. The baseboard for a large piece of background scenery that usually contains a LUMINAIRE strip.

cover. 1. To place a camera so that it can film all of an action area. **2.** To film a live event, usually for newscast purposes.

coverage. 1. The film taken of a news event; the filming of it. **2.** The number of counties in a geographical area that are covered by any level of a broadcast signal.

cover shot. 1. A wide-angle camera shot taken to provide protection for JUMP CUT close-ups when lip movements may not be in SYNC. **2.** Long shots to review the action within the action area.

cowcatcher. A sponsor's announcement that precedes the opening shots of the programmed material.

C.P. Construction Permit for a broadcast station issued by the Federal Communications Commission.

CPB. See CORPORATION FOR PUBLIC BROADCASTING.

crab dolly, crab. A support for a mobile camera that has steering controls enabling the DOLLY to move in any direction.

cradle head. A heavy camera mount with gears used to achieve tilting up and down; also used for panning shots.

crane, camera crane. A vehicle equipped with an oversized BOOM on which a camera can be mounted and thrust upward to a position higher than on a small DOLLY.

crane crew. Technicians who are responsible for the operations of the CRANE.

crane operator. The person who actually operates the camera CRANE, often one who drives the vehicle on which it is mounted.

crane shot. A shot made from a CRANE on which the camera is mounted.

craning. Moving the camera up or dropping it down by means of a flexible arm on the camera CRANE.

crawl title, crawling titles. See ROLL-UP TITLES.

credits, credit titles. The names of persons responsible for the artistic development of the film, such as actors, writers, directors, cinematographers, set and dress designers, sound engineers, etc., which appear on the screen either directly before the film is shown or directly following it.

creep. The slippage of videotape on the capstan (rotating spindle), which adversely affects playback synchronization.

creeper. An actor who moves too close to the camera or microphone.

creeper titles. See PULL-UP TITLES.

creepie-peepie. A hand-held television camera.

crew. The personnel who work on the production of a film other than the performers and those responsible for the creative process.

crew call. The typed notification to personnel of the time and place of shooting.

CRI (color reversal intermediate). A single-strand 16mm printing negative that is composed of sections taken from the original camera negative.

crisis. In a film drama, a moment of tension, a point at which the status quo is decisively changed.

crispening. Sharpening an image through digital picture information recirculation.

critical focus. A camera focus that is exact or a camera device that assures exact focus.

critics. Persons, usually employed by newspapers, magazines, or television stations, who evaluate motion pictures or television programs.

crop. To reduce or define the area in a shot, usually by tightening the frame.

cross. An actor's movement from one side of the set to the other in a camera shot, usually in a horizontal line before the camera.

cross-cut. To cut back and forth, from one action to another, usually between two scenes of different places in which the action is occurring simultaneously.

cross fade, cross dissolve. An audio source faded on one sound track while being raised to audio level on another track.

crossing the line. An awkward 180-degree shift in camera position, which is confusing to the viewer.

cross light, cross lighting. A technique of directing a light source on a subject at various angles to the axis of the camera lens.

crossover network. A circuit that divides electric impulse signals into their various frequencies.

cross-own. To own both a television station and a newspaper in a single market area.

cross plug. The mention of an alternate sponsor in a commercial.

crosstalk. Signal interference.

crowd noise. Muted conversation faintly heard in the background of a scene, used for realistic effect.

crowfoot. The metal brace on which a camera tripod is mounted.

crush. The intensification by electronic means of a picture's black and white levels.

crystal. A low-quality microphone in which the desired sound reproduction fails at approximately 8,000 Hz.

Crystal Award. Award given annually by the WOMEN IN FILMS organization to women for their contributions toward improving the image of women and encouraging the increased participation of women in the film industry.

crystal checker. A device that checks on the functioning of the crystal control on a camera or recorder.

crystal motor. A motor in which the speed is controlled by crystal vibrations.

crystal sync, crystal control. A system that controls the camera speed on the magnetic tape SYNC-PULSE signal through a scintillating crystal in both camera and audio recorder.

CS. A close shot of a performer on camera.

CTS (Communications Technology Satellite). A jointly owned United States and Canadian Hermes satellite employing solar energy, used since 1976.

CTW (Children's Television Workshop). A production company of public television which provides programming for children.

cue. Any sight or sound signal that prompts the beginning or end of action; a verbal signal to an actor who requires prompting.

cue card. An off-camera cardboard sheet on which are words or sentences as prompting devices for performers. Also called an *idiot sheet* or *idiot card*.

cueing. Any system, whether cue cards, hand signals, or audio signals, used to signal an actor or musician during his/her performance.

cue light. A small light near a narrator or performer that is switched on and off to indicate when he/she is to begin a line or paragraph.

cue mark. 1. A punched hole on film or sound track which aids in the synchronization of editing, projection, and recording by indicating the point at which to begin. **2.** A small white circle shown on several frames of a film reel just prior to its end; used as a warning to projectionists that a changeover from one reel to another is imminent.

cue patch. A piece of material placed on a film edge to activate a PRINTER LIGHT change or an automatic stop on the projection; it is made of adhesive magnetic or metallic substance.

cueprint. A positive projection print that contains the cues needed in POST-RECORDING.

cue sheet. 1. A sheet containing cues written in sequences to aid in audio mixing. **2.** A layout sheet used by animators.

cukaloris, cuke. See COOKIE.

cult film, cult movie. A motion picture that has an enthusiastic but limited following.

cume, cumulative audience. See REACH.

Curie point. The temperature at which the residual recorded signal of magnetic tape is reduced or lost.

curl. The bending of a film from the sides as a result of changes in humidity.

cursor. A moving dot on an electronic matting device, used for picture titling, which shows the letter-entry position.

curvature of field. A filmed image in which the focal points seem to lie on a curved rather than a flat plane.

cushion. The part of a program that can be lengthened or shortened according to the amount of time to be filled.

cut. 1. In *filming,* to change from one shot to another immediately. **2.** In *direction,* to stop action by the performers, camera, and audio equipment. **3.** In *film editing,* to eliminate unwanted portions, audio and/or visual, of a film. **4.** In *script editing,* to eliminate unwanted action or dialog in the typed script.

cutaway, cutaway shot. 1. An action shot that is not part of the principal action but is relevant to it and occurs simultaneously. **2.** The motion of the camera from the principal interest, the person being interviewed, to the interviewer.

cutback. A shot that returns to the principal action that preceded the cutaway. See CUTAWAY.

cut-in, cut-in shot. 1. A close-up shot of an otherwise indiscernible detail which is important to the action. For example, a love note, a map, or the face of a watch. **2.** Broadcast material inserted by a local station into network transmission.

cutter. The film editor who determines which scenes are to be deleted, which are to be kept, and in which sequence. **2.** A device for splitting a beam of light, usually a stick or GOBO edge placed in front of the light.

cutting copy. A positive print of a film on which editing can be done.

cutting in the camera. Shooting scenes or shots so that the desired content and continuity are obtained with little need for editing.

cutting on action. Shifting from one camera position to another, by using two or more cameras or using one camera for overlapping shots, to create a smooth and continuous flow of actors' movements. Also used to create tension or provide contrast.

cutting on the beat. Editing shots to begin at the start of a measure of music or on any beat within a measure.

cutting room. The room in which the film editor (CUTTER) works.

cutting sync, cutting in sync. The synchronization of the picture and sound track achieved by editing frame-by-frame with the use of a synchronizer.

cyan. A blue green element in color negative film.

cyclorama, cyc. A large, curved, plain (usually white) backdrop placed at the rear of a set or stage.

cyc strip, cyclorama strip. An illumination device that has several lamps in a row, used to evenly light a CYCLORAMA.

cycle animation. The photographed movement of animated characters by the repeated use of a series of CELS to show repetitive action such as running.

D

DA (directional antenna). A television antenna that picks up only specific frequencies.

DAC (dialog-to-analog converter). A device used to generate digitally transmitted signals to the original analog form.

dailies. Visual and audio WORKPRINTS of each day's shooting, shown that day or the next in order for the director and others involved in the filming process to judge the results.

daily production report. See PRODUCTION REPORT.

dark end. The area surrounding a film-processing machine which must have no illumination as the film is fed into it.

data rings. The rings on camera lenses that indicate the point of focus, depth of the area, and F-STOP.

day. A direction written in scripts and shooting schedules to indicate that filming is to be done during the day or, on set, to give the illusion of daylight.

day exterior. Referring to scenes or shots filmed out of doors during daylight hours.

day-for-night, day-for-night filming, D/N. A term denoting that shooting is to take place in daylight but will appear as at night, an effect achieved through special exposures, filtration, and processing.

daylight. 1. Referring to color film used for daylight shooting. **2.** Out-of-doors light that comes from the sun or sky.

daylight color film. Color film with a balance that provides good tones when it is exposed at color temperatures of approximately 5,500 degrees Kelvin.

daylight conversion filter, daylight filter. A camera lens filter that changes the color temperature of the light that passes through the lens, for outdoor exposure of color film which is balanced specifically for artificial light, usually 3,200 degrees Kelvin.

daylight loading spool. A light-tight metal spool that prevents light from touching the inner coils of film as it is being wound.

daylight projection. The projection of a film on a high-reflectance screen or onto a translucent screen from the rear or in a SHADOW BOX.

day part. The television broadcasting period during daylight hours when commercial advertising appears.

daytime. A classification for the sale of broadcast time, usually from 10:00 A.M. to 4:00 P.M.

day worker. A BIT PLAYER who is paid by the day rather than for the duration of filming on a motion picture.

DB (delayed broadcast). A film or videotape recording shown by local stations of a previously broadcast network program.

db meter. A device that indicates the levels of sound.

DBS (direct broadcasting satellite). A communication system to send multichannel programming directly from a SATELLITE (or satellites) into the sets of home viewers; will not be available until 1986 or later.

dead. 1. Referring to the acoustical quality of a confined area in which the reverberation is at a low level; **2.** equipment that is no longer operative; **3.** creative material that has been deleted, usually from a script.

dead air. A broadcast transmission that has no picture and/or audio signal.

deadline. The final date for the submission of material to be broadcast.

dead pan. With no expression of emotion; referring to an actor's face.

dead side. The side of a microphone that picks up responses less readily than the other side.

dead sync. See EDITING SYNC.

deaf aid. A small cueing earphone worn by a performer, usually a newscaster.

dealer spot. A live announcement added by a local retail outlet to a recorded television commercial.

dealer tie-in. A list of local retail merchants added to a network commercial for a product they carry.

deal letter. A letter by an advertising agency seeking a television production contract that discusses general contractual matters.

debug. To correct technical difficulties.

decamired. MIRED value of a filter or light source divided by ten, to obtain a more accurate color temperature rating.

decay rate. The rate at which an electronic signal or picture is diminished.

decibel, db. 1. A measure on a logarithmic scale of the relative intensity of sound. **2.** The ratio of two quantities of acoustical or electrical power indicated by a numerical value.

decibel meter, db meter. A device used to measure the intensity of sound against a random level of sound.

decisecond. One-tenth of a second.

deck. 1. The floor of the studio. **2.** An audiotape transport system without speakers or amplifiers.

decoder. The receiver circuitry in a color television set between the picture tube and the signal detector.

decor. Set or location furnishings used in a film.

decorative properties. Properties used on a set or location to provide atmosphere and an appearance of authenticity; not to be confused with PROPS, which are objects actually used by the actors.

deep-field cinematography, deep-focus cinematography. CINEMATOGRAPHY that brings into sharp focus objects at various ranges, both close and far, with the use of small F-STOPS and/or short FOCAL-LENGTH lenses.

definition. 1. The rendition of the intricate details of an image. **2.** The clarity of sound or image.

defocus. To deliberately take the action out of focus by focusing the lens to a close point in order to reduce the depth of the action area; used for special effect.

degauss. 1. To erase a recorded tape by realigning all particles in a regular pattern. **2.** To eliminate any unwanted magnetism from metallic equipment used for editing, recording, or playback.

degausser. Any device that erases previously recorded material on magnetic tape or film, or demagnetizes magnetic recording heads.

degradation. A loss of image fidelity through DUPLICATION.

demagnetize. To DEGAUSS the ability of some pieces of metal to produce unwanted magnetic fields, particularly that developed in tape-recording heads.

demagnetizer. An electromagnetic tool used to DEGAUSS tape-recorder HEADS.

demodulate. To withdraw a broadcast signal from a CARRIER wave.

demographic profile. A statistical study of television audiences divided into categories such as age, sex, education, and economic levels.

denouement. The action in a film which follows the CLIMAX and ties up all loose ends that may remain after the main conflict is resolved.

densitometer. A measuring device used to denote film DENSITY.

densitometry. The technique of measuring the density of filmed objects.

density. The degree of OPACITY in photographic film.

depth of field, D/F. The distance range from close to far in which the camera retains sharp clarity of focus.

depth-of-field scale. A numbered indication on a camera lens which shows the depth of the field at specific distance and aperture settings.

depth-of-field table. A chart that lists the DEPTH OF FIELD for a lens of designated focal length at different APERTURE and FOCAL POINTS.

depth of focus. The distance between the camera lens and the film in which focus retains sharp clarity.

derived sound. Sound that is removed from two STEREO sound tracks and transferred to a third loudspeaker.

desaturation. The removal of color from film to achieve a MONOCHROME result.

detail shot. A close shot used to reveal details of an object or part of an object, or of part of an actor's body (such as a hand with a missing finger).

deuce. A floodlight with a 2,000-watt lamp.

deus ex machina. A plot device used in a film to provide a too-easy solution to the protagonist's problems, one inconsistent with realistic probabilities. (Latin. "God from a machine.")

develop. To produce a latent image on exposed film by treating it with chemicals and light-sensitive materials.

developer. A chemical liquid used to DEVELOP film.

development. 1. The process of chemically producing a permanent image from a latent image on film. **2.** Sequential changes in the plot and/or an actor's character that enhance a film's dramatic effect.

DGA (Directors Guild of America). An organization of professional film and television directors.

diagonal cut, diagonal splice. The process of joining two pieces of film at an oblique angle; used with magnetic film to reduce the noise of the cut or splice.

dialog. Conversation between two or more actors.

dialog coach. A person who helps actors rehearse lines and instructs them on how to improve required accents, delivery, and character interpretation.

dialog replacement. The process of DUBBING or LOOPING additional or improved dialog on a dialog track.

dialog track. A sound track that carries lip-synchronous dialog (or monolog), as compared with sound tracks that carry music and other sound effects.

diaphragm. An aperture with overlapping metal leaves in a camera lens (occasionally in a spotlight) which controls the amount of light passing through the lens.

diaphragm presetting. The selection of F-STOP to achieve the desired DEPTH OF FIELD so the action can be lighted to the level required for correct exposure.

diapositive plate. A glass plate containing a photographic transparency placed in front of a camera to add a visual component to the scene being shot.

diary homes. The 25,000 homes in which participants in a NIELSEN research study log their television habits and provide demographic information such as age, sex, and economic status of the persons watching the programs.

dichroic. Referring to the process whereby layered coatings on filters, mirrors, and reflectors control the spectral characteristics of light; used to change the color temperature of lights.

dichronic mirror. A color filter that separates red, green, and blue light elements in a camera for the camera pickup tubes.

differential rewind. A device used to achieve the simultaneous winding of film from two or more reels with different diameters.

diffused light. Light dispersed either by conditions in the atmosphere or by use of LUMINAIRES, so that no shadows or muted shadows are cast.

diffuser. 1. A lens attachment that reduces the image sharpness in part or all of an action field. **2.** Any translucent or transparent material used in lighting to distribute the light and thereby soften the lighting on the subject.

diffusion. The dispersion of light by use of a translucent or transparent DIFFUSER.

diffusion screen. A translucent screen used for the even diffusion of light, particularly in LIGHT BOXES, used for creating titles or for filming three-dimensional objects without shadows.

digital counter. The footage indicator on a tape recorder.

dilute. To reduce color saturation by adding white.

dimmer. A device used to control the light output of LUMINAIRES.

dimmer bank. A cluster of dimmers, sometimes installed on a mobile unit such as a track or cart.

dinky, inky. See INKY DINK.

diode. A vacuum tube that contains negative filament and positive plate elements.

diopter, diopter lens. A lens attachment used to shorten the focal length for close shots.

diplexer. Equipment that allows both audio and visual signals to be transmitted from the same antenna.

dipole. FM antenna.

direct cinema. Nonfictional cinematography in which hand-held cameras and sound recorders film "on the spot" action with natural sound. Cf. CINÉMA VÉRITÉ.

direct color print. A one-step color print made from the original film.

direct cut. The linking of two shots in a scene to preserve continuity in time and action.

direct sound. See INDIGENOUS SOUND.

directional microphone. A microphone with a greater pickup intensity in one direction than in others; used in filming to reduce pickup of sound not included in the principal action.

directional pattern. The manner in which a microphone picks up sound, whether nondirectional, bidirectional, or unidirectional.

direction of glance, direction of look. The direction in which an actor looks as the shot is finished.

director. The person responsible for the quality of the film as it appears on the screen, having had control of the cinematic techniques, the actors' performances, the credibility and continuity, and the dramatic elements of the production.

director of photography. See CINEMATOGRAPHER.

director's finder. A small calibrated device with a variable angle of view used to view the ACTION FIELD in order to find the proper lens focal length for the shot.

direct positive. See REVERSAL.

direct response. Referring to broadcast advertising that requests the audience to purchase the product by telephone or mail order within a limited period of time.

dirty dupe. A single-light single-strand duplication of a WORKPRINT.

discontinuity. The effect of anything that occurs as one shot ends and another begins which robs the second of complete continuity with the first, whether in lighting, props, furnishing, objects, or position of actors.

discovery dolly shot, discovery pan, discovery zoom. A shot that exposes something to the audience that was not in view when the shot began.

dish, dishpan. A large concave antenna, usually identified with cable television.

disk. 1. Equipment used to achieve SLOW MOTION or FREEZE-FRAME shots on videotape. **2.** A technique that uses LASER beams and metal styli for the production of large quantities of television recordings. **3.** A phonograph recording.

disk pack. A set of disks in which recorded information is stored.

disk recorder. A video recorder that uses one magnetizable disk or DISK PACK to store information.

displacement magazine. A film magazine that has such tightly packed SPINDLES that a film load intrudes on the space it will occupy after exposure.

display. A visual review of stored information.

dissolve. The linking of two shots as the first fades out and the second fades in, usually signifying a lapse of time and a change in place.

dissolve animation. The achievement of slow changes in animation objects or artwork by using quick DISSOLVES.

distant signal. The process by which cable programs are distributed in localities beyond the system's usual reception area.

distortion. 1. A misshapen image caused by a malfunction of the optical system. **2.** A departure from the norm of a SIGNAL wave or FREQUENCY.

distortion optics. Special-image effects created by optical attachments used on a camera or printer.

distribute. To send an electric signal along a specific course.

distribution. The process of releasing motion pictures through sale, lease, or rental agreements to distributors, such as motion-picture theater owners.

distribution amplifier. An electronic device used to send television signals to MONITORS at the original level without loss of sound.

distributor. Any organization that circulates motion pictures through sale, lease, or rental agreements or, in the case of libraries, loans.

ditty bag. A canvas bag containing small instruments or items needed by a camera crew; sometimes swung beneath a TRIPOD.

divergent lens. Three kinds of camera lenses capable of forming negative images: plano-concave, bi-concave, and divergent-minicus.

divergent turret. A camera turret, larger than a flat turret, which holds lenses so that their axes spread out to prevent long-focal-length lenses from covering the same ACTION FIELD that short-focal-length lenses cover.

DMA (Designated Market Area). A classification of the A. A. Nielsen Company research survey, which shows the cluster of counties in which home viewers watch the local television channel.

D-max. The density in a REVERSAL film of the totally unexposed areas of an EMULSION following development and fixation.

D-min. The density of a REVERSAL film of the totally exposed areas of an EMULSION following development and fixation.

dock. See BAY.

docudrama. A television dramatic film based on historic or current events in which factual data are emphasized: e.g., *The Day the Bubble Burst,* depicting events that led to the stock-market crash in October 1929.

documentary, documentary film. A motion picture depicting nonfictional events or occurrences, filmed on-the-spot and presented from a particular point of view, meant to be informative or make a specific comment on a subject or issue.

dog. A badly flawed production.

Dolby system. A method of reducing the noise level of a tape or optical recording by the manipulation of the high-frequency response.

dolly. 1. A small camera-and-operator mount used for mobility during action filming. **2.** To move the mount during the actual filming.

dolly pusher. Any member of the camera crew who moves a DOLLY prior to, or during, the filming of a shot.

dolly shot. A shot made from a moving DOLLY.

dolly tracks. Rails on which the DOLLY is moved during the shot.

dope sheet. See EXPOSURE SHEET.

dot. A small round sheet of opaque material used to block direct light from an area on which it might normally fall.

double. A person who looks like the star performer and substitutes for him/her when physically dangerous shots are to be made, or who can substitute in long shots when the actor's features are not shown clearly, or in closer shots when only the star's back is shown.

double broad. A box-shaped 4,000-watt light used as a FILL LIGHT.

double chain. To run A AND B film rolls on two projectors in order to achieve CUTAWAY shots during a television interview.

double clad. A backdrop FLAT on both sides of which scenery is painted.

double exposure. The recording on a single filmstrip of two images superimposed or shown side by side.

double feature. Two motion pictures on the same program, shown for the price of one.

double headset. An intercom system in which there is a separate circuit for each headphone.

double-perforation stock. Film with perforations along both edges.

doubleprint titles. Screen titles that are superimposed over action or images by the A AND B ROLL PRINTING process.

double scrim. Two pieces of SCRIM (usually gauze) pressed together to produce a darkened image.

double-system sound recording. The sound recording on magnetic tape or film which is separate from the camera but synchronized with it.

double take. 1. An actor's delayed reaction to something he/she has just seen, causing him/her to take another startled look. **2.** Action shot from different angles and edited in overlaps by mistake.

down. A film script instruction to reduce the level of sound, particularly musical sound.

downgrade. To lower the status of a performer.

downlight. A LUMINAIRE that is beamed directly above the subject.

down link. The transmission from a SATELLITE to an earth station.

down-shot. See HIGH-ANGLE SHOT.

down stage. That area of a stage which is nearest the camera or audience.

down the line. Movement toward the destination of an electric impulse.

dowser. An electric device used in motion-picture theaters to cut off a projector's light beam while the changeover between reels is being made.

drain. The loss of power in a storage battery.

drama. The performance by actors of roles written in a script for presentation to an audience. (From the Greek *dran,* "to do.")

dramatic film. A motion picture that shows the protagonist engaged in extraordinary conflict, usually with strong emotional impact arising from crises and climax.

dramatic irony. An awareness by the audience of information not available to actors, a device used to trigger and heighten suspense.

dramatic unity. The principle of confining action and events to a limited area and time and presenting them through tightly interwoven relationships.

drape. A background made of plain, unpainted, heavy material.

draw a blank. To have a momentary lapse of memory, as when a performer forgets the next lines in a script.

draw cards. Credits or titles written on cards that are placed in a special holder for easy and accurate access before the camera (but out of camera range).

dream balloon. An image that appears near an actor's head to indicate the content of his/her daydreams, visions, or inner dialog.

dream mode. Shots used to reveal what is occurring in a performer's imagination.

dress. To prepare a set with furnishings, artwork, and props prior to SHOOTING.

dresser. **1.** An assistant to the WARDROBE MASTER/MISTRESS. **2.** An actor's personal wardrobe assistant.

drift. The tendency of an electronic circuit to react to changes in time and temperature.

drive-in, drive-in theater. An outdoor motion-picture theater equipped with parking spaces for automobiles in which the audience remains seated for the performance; each parking space is equipped with a loudspeaker for placing inside the car.

drop. A heavy canvas on which a scene is painted; used as a background for action shots.

dropout. A horizontal PLAYBACK streak across a picture caused by irregularities on the tape surface during recording.

drop shadows. Shadows used with title lettering to sharpen the outlines of the letters.

drum. **1.** A rotating holder for film slides. **2.** A rotating mount for vertical strips of titles and credits. **3.** A flywheel used to facilitate the smooth flow of film past the projector SOUND HEAD.

dry box. The last unit in a film-processing machine which circulates warm air around the developed film.

dry cell. A storage battery without water.

dry end. The section of a film-processing machine that contains the DRY BOX.

drying. Circulating dry warm air around developed film in the final stage of processing.

dry-mounting. Pressing artwork to mountboards by using heat and a sheet of dry-mounting tissue.

dry-mounting press. A press with two flat PLATENS used to melt dry-mounting tissue that has been inserted between the artwork and the mountboard.

dry run. A rehearsal of actors or a run-through of camera movements prior to the shooting of a scene or shot.

dub. 1. To record dialog, with lip synchronization, to be added to a film after it is completed. **2.** To insert English dialog (recorded) in a film in which the actors speak another language. **3.** To duplicate audio or videotape masters.

dubber. Audio playback equipment with magnetic sprocketed film.

dubbing. The transposition of an electric impulse or signal from one tape recorder to another.

dubbing session. A session in which performers record dialog while they view a WORKPRINT of the action that accompanies their speech. Recording by actors who are speaking in a language other than that of the original film.

dub off. To record all or portions of sound from another recording.

dulling spray. A liquid wax-based aerosol spray used to reduce the reflection on shiny surfaces.

dump tanks. Large water-filled tanks perched at the top of steep chutes, which, when released on cue, will cause the water to rush violently downward creating the illusion of a tidal wave or flood.

duopoly. The ownership of two or more broadcast facilities in one market area.

dupe. 1. A copy of a film or tape recording. **2.** The process of making such a copy. **3.** To make a DUPLICATE NEGATIVE from a positive film.

duplicate negative. 1. A negative made from a master positive or a positive film to be used for making RELEASE PRINTS. **2.** A negative copied from the original negative.

duplication. Receiving the same program or programs on a home television set more than once.

dutch angle, dutch tilt angle. A camera angle made by tilting the camera in any direction other than a vertical or horizontal position.

dutchman. A strip of canvas used to cover the hinges between FLATS in order to conceal the cracks.

DX. See BIPACK.

dye coupler. Any chemical element added to film EMULSION or developer.

Dynalens. The trade name for a liquid optical mount that reduces the vibrations in the camera during shots made from a moving vehicle; also used with telephoto lenses.

dynamic, dynamic microphone. A microphone sensitive to pressure and whose diaphragm is attached to a moving coil in the magnetic field.

dynamic cutting, dynamic editing. The meticulous joining of related shots that were not made in sequence.

dynamic duplication. An AC transfer system that achieves duplication through the use of a videotape master and several SLAVES.

dynamic range. The variance between the lowest and loudest sounds in a live sound or recording pickup.

E

Early Bird. The original name for an eighty-five-pound commercial GEOSYNCHRONOUS SATELLITE.

earphone, earpiece. A small receiver worn in the ear, usually by newscasters.

earth station. A structure located on the ground, used for the reception of SATELLITE transmissions.

easel. A stand on which are placed graphic arts materials to be included in an action shot or photographed for other uses.

east. The right side of a table used for ANIMATION production.

EBR (electron beam recording). A videotape-to-film transfer system in which STEP PRINTING of three black and white negatives is used.

EBS (Emergency Broadcast System). The warning system for use by the United States government in case of a national emergency, during which all broadcast stations are electronically placed under federal authority.

echo. The (usually) unwanted duplication that either follows or precedes a primary signal.

echo chamber. 1. A room or other boxed enclosure in which reverberation is added to sound. **2.** A device, either electronic or acoustical, which adds slight reverberation to an audio signal.

edge fog. A hazy unwanted area that appears along the edge of a portion of film, usually at either end of the roll; caused by the presence of light during loading or unloading of film spools.

edge numbering machine. A mechanism used to print the EDGE NUMBERS on processed film and magnetic film sound tracks.

edge numbers. The manufacturer's identification numbers that appear on the FILM STOCK, used when matching negatives while editing the WORKPRINT. These numbers may be ink-printed for all relevant materials so that the audio and visual rolls can be correlated easily.

edge track. An audio band on the edge of magnetic film opposite the perforated edge on a SINGLE-PERF roll.

edit. To correlate, arrange, synchronize, trim, or cut film, and to annex leaders to it and/or sound track STRIPS in order to achieve the properties and proportions necessary for a cohesive and credible film production.

edit code. A videotape retrieval system in which data concerning the frames, seconds, minutes, and hours are recorded on a control track.

edited music track. Any background music or sound that is correlated with the foreground action.

editing. The creative process of correlating, rearranging, synchronizing and cutting the film, both audio and visual, to produce the desired final version.

editing machine. A machine in which the film moves in one of two ways, either left to right from the FEED PLATE or down from the feed plate; permits the editor to view the FILMSTRIP in short lengths and to control its speed and its backward or forward movement by a foot pedal.

editing ratio. The footage ratio, usually about seven-to-one, between the exposed film stock and the final edited version of the film.

editing sync, editorial synchronization, dead sync. The synchronization of the WORKPRINT and SOUND TRACK, frame by frame, without allowance for film pull-up delay.

editor. The person responsible for making a coherent whole of the various audio and visual components through splicing and cutting, (re)assembling and (re)organizing the SHOTS and SOUND TRACKS.

editorial assistant. Assistant to the editor.

editorial process. The ways and means in which a film is assembled by the editor.

edit out. To delete sections of a visual film or sound track.

educational film. Any film made for purposes of instruction or the dissemination of information; often used in classes.

educational television. Programs presented for purposes of instruction or for intellectual enlightenment, not supported by commercial advertising.

effect filter. Any FILTER used on a camera lens to achieve or provide an additional element to the scene being filmed, such as haziness over a scene in a steamy tropical jungle.

effective aperture. The ratio between the focal length and width of a lens APERTURE in which the image is clearest; depends upon the IRIS diaphragm.

effects. 1. Electronically produced visuals. **2.** Any elements of sight or sound used for specific effects.
effects bank. A control-room console from which the electronic optical effects are transmitted.
effects projector. A film projector used to project visuals on a translucent screen or other backdrop.
effects track. A sound track that contains SOUND EFFECTS, such as a train whistle or roaring waterfall, but does not contain either voices or music.
EFP (electronic field production). The use of portable videotaping equipment when filming on location.
eightball. A small sound microphone.
eight-millimeter film. Film that is 8mm wide, having smaller perforations than the older film stock, which had perforations the same size as 16mm film (though twice as many).
electric-drive motor. Any electric motor that is used to motivate the mechanism that runs a camera, projector, or recording device.
electrician. The crew member who has charge of the lighting equipment on the set.
electrode. The terminal of an electric source.
electrolysis. An electrical charge that occurs when electric current is passed through an ELECTROLYTE.
electrolyte. An ionized liquid or paste substance that is a conductor for electric current.
electromagnet. A soft iron core that is magnetized by an electric current passing through the coil in which it is contained.
electromagnetic radiation. The passing of radiant energy through matter or space in invisible waves.
electron beam recording. A direct-transfer technique of high-quality videotape to film, involving the use of STEP PRINTING of three black and white negatives.
electronic clapper. A camera device that simultaneously flashes a few lengths of film and equal lengths of sound to a tape recorder, so that the visuals and sound can be synchronized for WORKPRINTS.
electronic composite. A sound RELEASE PRINT on which the sound track was taken from a master MAGNETIC TRACK.
electronic editing. A recording of original videotape signals on a second videotape, including changes to eliminate the need for splicing.
electronic still store. A machine that stores still pictures, tapes, frames, and graphic arts designs in magnetic packs.
electroprint. See ELECTRONIC COMPOSITE.
electroprinting. A technique used in making RELEASE PRINTS in which the sound is fed directly from the master MAGNETIC TRACK to the release print without requiring an optical sound track printing master.
electrostatic microphone. See CONDENSER MICROPHONE.

elevation, elevation drawing. A scale drawing that shows the shapes and measurements of the vertical surfaces of a set.

elevation shot. A shot in which the camera moves in a vertical position only.

ellipsoidal spot. A spotlight that produces a sharp beam through the use of SPHERICAL OPTICS.

elliptical cutting, elliptical editing. A characteristic of film editing in which portions of the action are eliminated, usually to increase the tempo of the action and usually accomplished through the use of JUMP CUTS.

Emmy. An annual award presented for the most outstanding creative contributions in each of many fields (acting, writing, directing, cinematography, set designing, etc.) by the NATIONAL ACADEMY OF TELEVISION ARTS AND SCIENCES.

emote. To project a variety of reactions (anger, despair, joy) in order to meet the requirements of a film role.

empathy. The subjective relation of the audience to the filmed performances, such as feeling frightened when the character is in danger.

emulsion. The light-sensitive coating placed on cinematographic film by coating or bonding. Gelatin and silver salts are used on unprocessed film, gelatin and metallic silver on processed film, and iron oxide on magnetic sound film.

emulsion pile-up. A tiny collection of EMULSION in the GATE of a projector, camera, or viewer.

emulsion position. The EMULSION side of a film as it is placed in a projector; if it is toward the screen it is in a B-WIND projection position; if it is toward the projection lamp it is in an A-WIND projection position.

emulsion side. The side of the film that bears the EMULSION coating.

encoder. A mechanism that changes the characteristics of an electronic signal or adds information to it by superimposition.

end rate. The least expensive commercial advertising time rate offered by a broadcast station.

ENG (electronic news gathering). The making of a news program with hand-held cameras and video cassette recorders.

enhancer. A signal processor used to achieve a sharper or more clearly defined picture.

epic. A film in which the leading actor is cast in a role of heroic dimension, performing acts of derring-do in a drama of considerable range; usually a costume film.

episodic. Referring to a film in which the emphasis is placed on incidents rather than on the subtleties of plot or characterization.

EQ. See FREQUENCY EQUALIZER.

equalize. To adjust electronically the frequency and intensity qualities in a sound or picture source.

Equal Time. A section of the Federal Communications Act (1934) that guarantees equal broadcast time privileges to all major candidates who run for the same political office in an election.

erase. To eliminate or neutralize recorded electromagnetic signal patterns before recording.

erase head. A degaussing device that eliminates recorded signals by reassembling the magnetic particles in an even pattern. See DEGAUSS.

establishing shot. A long shot that establishes the primary locale of the film; used for atmospheric purposes and generally shown in the opening scene of a film.

estimate. The tentative budget requirements for a film production.

ETV (educational television). The presentation of scholastic programs on a noncommercial basis.

exciter light. The light that throws a beam on a photographic sound track.

executive producer. The person in charge of the financial affairs of a motion-picture production; he may also be actively involved in the creative processes of the production, occasionally being both executive producer and PRODUCER.

exhibitor. The owner or operator of a motion-picture theater.

existentialist films. Loosely, the American detective films of the late thirties and the forties in which a man of no particular significance meets and defeats the forces of evil in head-on confrontation. Found a following among French existentialists of that period. Among these films were *The Maltese Falcon* and *The Big Sleep,* made from novels by Dashiell Hammett and Raymond Chandler respectively. The genre is better known in France as *film noir* (black film.)

exploitation films. Motion pictures that are intended to appeal to a specific and limited audience, such as horror films or films starring rock musicians; usually low-budget films. Also, pornographic films.

exposition. Disclosure usually through dialog of past events or facts that are vital to the plot as it unfolds.

exposure. The subjection of raw film stock to the action of light in order to achieve a LATENT IMAGE.

exposure guide. Data that show the best exposures for particular film in a variety of lighting situations.

exposure index. Numbers placed on film by manufacturers, indicating the EMULSION speed to be used under specified conditions.

exposure latitude. The degree to which film can be underexposed or overexposed and still present a clear image.

exposure meter. A device used to gauge the amount of direct or reflected light to be given to a film for the best results. Also known as *light meter.*

exposure rating. See EXPOSURE INDEX.

exposure sheet. Information or cue sheet used in filming an ANIMATION sequence.

expressionism. In films, a presentation of the world, not realistically, but as the character (actor) believes it to be; sometimes indicated by bizarre distortions in the sets, scenery, costumes, and makeup.

extension tube. A tube placed between the camera and the lens which can extend the focal length of the lens and improve its power of magnification; used in close-up photography.

exterior (EXT). 1. A film script term indicating that the action is to take place outdoors. 2. Referring to any action shot outdoors.

exterior lighting. Any lighting, artificial or natural, used on outdoor locations.

exteriors. Shots made on outdoor locations or on sets in which the action seems to take place outdoors.

extra. An actor who plays a very minor role in a film production; usually has only a few lines or none. Often appears in crowd scenes.

extreme close-up (ECU). A shot made of a small object or some detail of an actor's face, hands, etc.

extreme high-angle shot. A shot made from a down-tilted camera placed high above the action.

extreme long shot (ELS). A shot that encompasses a considerable distance, such as a view across a mesa of a stagecoach appearing on the horizon.

extreme low-angle shot. A shot made from an upward-tilted camera placed below the action.

eye contact. A performer's direct look into the camera.

eye-level angle (shot). A shot made from the eye level of the performer in order to present the scene as it appears to him/her.

eye light. A light without great intensity that is placed near the camera and used to highlight the performer's eyes.

eyepiece lens. The lens on a camera VIEWFINDER against which the cameraman places his eye.

F

f. A symbol that denotes the relationship between the camera lens opening and the focal length.

faceplate. The front surface of a television picture tube.

facilities (fax). The equipment used by technicians in the production of a film: e.g., cameras, LUMINAIRES, and microphones.

fade. 1. To slowly lower the volume of sound or music that accompanies a film. **2.** Sometimes used to mean DISSOLVE.

fade-in. 1. The process in which the filmed picture emerges from a darkened screen to reveal a fully lighted image, traditionally used on the motion picture screen to open the film. On television it is commonly used to open each act. **2.** A gradual increase in sound, from inaudible to audible.

fade-out. 1. The process in which the screen gradually darkens to black, finally fading out the entire picture, traditionally used on the motion-picture screen to end the film. On television, fade-outs must end each act that has been faded-in. **2.** A gradual decrease in sound from audible to inaudible.

fader. Any device that raises or lowers the audio levels, or darkens or brightens the visual images.

fades. See FADE-IN, FADE-OUT.

fade scale. A FADER device used to aid in making smooth transitions in the increase or reduction of exposure.

fading. A gradual loss or wavering in broadcast signals caused by changes in time or in atmospheric conditions.

Fairness Doctrine. A mandate by the Federal Communications Commission that obliges television stations to make broadcast time available to opposing viewpoints on issues of public concern. Upheld by the United States Supreme Court in 1969. (In September 1981, a majority of the FCC voted to urge the repeal of the Fairness Doctrine, claiming that the EQUAL TIME provision amounts to government censorship. The issue must now be voted on by Congress.)

fall-off. The reduction in the degree of light as the distance increases from the LUMINAIRE source point.

false move. An action or movement by an actor which is untrue to the character he/she portrays.

false reverse, false reverse angle shot. A reverse-angle shot made when the camera position changes radically so that the performer is out of place in the edited shots. See REVERSE ANGLE.

Family Viewing Time. The hour of viewing time from 8:00 to 9:00 P.M. during which the network programs downplay violence and overt sex.

fan. Any person who is an ardent, persistent, and perhaps uncritical admirer of a performer and/or a particular film or film genre.

fantasy film. A film portraying unreality, in which the story is concerned with the bizarre and/with fanciful persons and events: e.g., the early Disney films such as *Snow White and the Seven Dwarfs* and the recent *Superman II.*

farce. A comedy of broad humor.

far shot. A long shot.

fast. Referring to **1.** an emulsion that is extremely sensitive to light; **2.** camera lenses with an f-value which transmit a maximum degree of light.

fast film. A film with a high sensitivity to light.

fast forward. The rapid speed at which tape is moved from the feed to the take-up reel.

fast lens. A lens capable of collecting a great amount of light.

fast motion. See ACCELERATED MOTION.

favor. 1. To give one performer more audience exposure than the others in a shot by the use of camera angles or favorable lighting. **2.** To give preference to a performer by training the microphones or camera on him/her.

fax. See FACILITIES.

FCC (Federal Communications Commission). An agency of the United States government established by the Federal Communications Act (1934) to oversee the nation's broadcasting.

feature. 1. To present a performer in a principal role in a motion-picture or television film. **2.** FEATURE FILM.

feature film. A full-length film production (usually at least ninety minutes) meant to be commercially exhibited in motion-picture theaters.

feature player. An actor who plays important supporting roles in films but is never the star performer.

Federal Trade Commission. See FTC.

feed. 1. To transmit a broadcast signal. **2.** The mechanical process of guiding film through a camera, printer, projector, or processor.

feedback. The input signal returned from an outgoing signal in a specially arranged electronic circuit.

feed lines. To read or speak lines of dialog, out of camera view, to actors who are being filmed; used to cue an actor on forgotten dialog, either by a person off-stage or by another on-stage actor.

feed plate. A spinning plate on an editing machine from which the film is fed through the sprocketed optical path to the TAKE-UP PLATE.

feed reel. A supply reel for a film projector or tape recorder from which film or tape is pulled.

feed spool. A spool from which film is fed into the film projector.

feet. Standard measurement of film; sixteen 35mm, forty 16mm, and seventy-two Super 8mm per foot of film.

female. An electric connection plug with hollow sockets which receives a corresponding, or MALE, plug.

femtosecond. One-quadrillionth of a second.

FET (field effect transistor). A SEMICONDUCTOR used for amplification.

fiberglass. Thick plastic sheets used to create scenic props that are large but light in weight.

fiber optics. A laser beam wave guide television system that is interference-proof and can transmit 167 channels on one 90-micron flexible glass fiber cable.

fidelity. Accuracy in reproduction of an original sound signal; also the accurate reproduction of color.

field. 1. ACTION FIELD. **2.** Half a television picture scanning cycle with two alternate scan-line fields to each frame, or sixty fields per second.

field angle. An angle that contains 90 degrees of the illumination from a spotlight.

field camera. A portable camera used for filming on location when mobility is required.

field frequency. The number (60) of transmitted television fields per second.

field guide. A printed chart, used in ANIMATION work, which shows the areas to be filmed from several different camera positions. Made on a sheet of transparent acetate equipped with animation peg holes.

fifteen ips. The recording speed of audiotape.

fill. Any material used to lengthen a program if the prepared material is running short.

fill leader. A length of blank film substituted in WORKPRINTS for a damaged or accidentally blank section of film; used in sound to connect two sections of the workprint.

fill light, filler, fill-in light. 1. A LUMINAIRE used to supply overall illumination in order to obscure shadows created by a brighter KEY LIGHT. **2.** The illumination itself.

film. 1. Sprocket-holed rolls of flexible transparent cellulose triacetate coated with various light-sensitive EMULSIONS and iron oxide. **2.** To photograph a motion picture. **3.** The motion picture itself. **4.** Generally, the cinema. **5.** To make a motion picture.

film aesthetics. The intellectual approach to cinema in which the methods of obtaining a particular audience response are studied.

film archive. A place where motion pictures are stored and are (usually) available for research by serious students of film.

film base. The filmstrip base that carries the EMULSION and iron oxide for photographic purposes.

film chain. All equipment, such as camera, cables, power supply, and monitor, through which motion pictures are projected in a television system.

film chamber. A container that is lightproof and holds both exposed and unexposed film, such as the film box on a camera.

film checker. The person who examines film for possible defects.

film-cleaning machine. A machine that cleans film by pulling it through a solvent-dipped cloth or through an ultrasonic cleaning mechanism.

film clip. See CLIP.

film commissions. See STATE FILM COMMISSIONS.

film criticism. Opinions concerning motion pictures expressed by professional reviewers for the print or electronic media or by authors of books concerned with the aesthetics of films and filmmaking.

film cutter, film editor. See EDITOR.

Filmex. The name by which the Los Angeles International Exposition is commonly known; held for two weeks annually, this event screens dozens of domestic and foreign films which compete in various categories including features, first films, documentaries, and short films.

film festival. 1. The exhibition of many competing films during a short period of time, for the purpose of choosing the best in several categories for special awards. **2.** A period at a theater during which a special kind of film or films starring a particular performer are shown.

film gate. The components of the APERTURE PLATE and the PRESSURE PLATE in a camera or projector.

film gauge. The width of various types of motion-picture film.

film handler's gloves. Thin cotton gloves worn by any person who handles original film to prevent fingerprints from smudging it.

filmic. A term referring to things that take place only in motion pictures, as opposed to CINEMATIC, which can apply to almost anything connected with filmmaking. Something that appears *on the screen,* such as one scene cutting to another, is filmic.

filmic time and space. The manner in which time and space are treated in a motion picture through the use of FLASHBACKS and FLASH FORWARDS and CUTS and DISSOLVES, which carry the audience instantly from one time and place to another.

film island. The group of film and slide projectors in a television studio.

film leader. See LEADER.

film library. See FILM ARCHIVE.

film loader. The member of a camera crew who is responsible for placing film rolls in the magazine of the camera.

film loop. 1. A small section of slack film placed between sprocket rollers and intermittents to prevent tearing the film. **2.** A long spliced length of film or sound track which can be run through a PRINTER or sound playback film MAGAZINE. **3.** A lightproof container from which unexposed film is fed into a camera and which also contains, in another compartment, the roll of exposed film. **4.** The lightproof container that holds film to be fed into the processing machine.

filmmaker. A person who makes motion pictures, particularly one who is heavily involved in, and largely responsible for, the production process.

film noir. Films that deal almost exclusively with violence and crime. See EXISTENTIALIST FILMS.

filmograph. A film of still pictures as opposed to motion pictures.

filmography. Pertaining to the study of, or writing about, motion pictures.

filmology. The study of films and/or filmmaking.

film perforations. See PERFORATIONS.

film phonograph. A mechanism that reproduces sound from an OPTICAL SOUND TRACK to be used for rerecording or for playback purposes.

film plane. The location of the front surface of film in relation to the lens.

film review. A critical analysis of a film by a critic, either given orally on television or radio for a program audience or written for magazine or newspaper. See FILM CRITICISM.

film running speed. The rate in frames per second or meters per minute at which film runs through a camera or projector.

film splicer. See SPLICER.

film stock. Unprocessed film or any particular type of film.

film storage. The safekeeping in containers of film stock and all other film (exposed or not) until it is needed. Original film is usually stored in vaults where the temperature and humidity are controlled.

filmstrip. A section of 35mm, 16mm, or 8mm film for showing on a special projector one FRAME at a time.

film structure. The presentation of a film's dramatic action in order to reveal past, present, and future.

film talent agency. A professional group that represents performers in deals made with studios or other employers, for a percentage of the performer's salary.

film transfer. See TRANSFER.

film transport. See TRANSPORT.

film treatment. 1. The elimination of unwanted moisture from processed film, before it is lubricated. **2.** The restoration of film that has been damaged. **3.** TREATMENT.

filter. 1. A lens element of optically ground glass that absorbs specific visual components and rejects others. **2.** A sound-equipment device that attenuates specific narrow frequency bands. **3.** A mechanical and/or chemical device used to remove sludge from processing solutions.

filter factor. A number that indicates the degree of light absorption admitted through a filter into the optical system.

filter holder. A device used to hold a filter before a lens or in a printer or camera slot.

filter slot. A groove in a printer or camera in which a FILTER HOLDER is placed in order to securely position the filter before the film.

filter wheel. A FILTER HOLDER that is attached to a position behind the camera lens.

final cut. 1. An editor's last work on film WORKPRINTS before they are brought into harmony. **2.** On a sound print, the editor's last workprint before the sound is mixed. **3.** The composite print that results from the combination of action film and sound track.

final shooting script. See SHOOTING SCRIPT.

final trial composite. The film that is the last, and approved, version of all previous trial COMPOSITES.

final trial print. See FIRST ANSWER PRINT.

finder. See VIEWFINDER.

fine cut. A WORKPRINT that has been meticulously edited so that usually no further editing is required.

fine grain. A term applied to a film EMULSION that contains minute silver particles and a more transparent base than usual.

fine grain duplicate negative. A DUPLICATE NEGATIVE made from a MASTER POSITIVE which had itself been made from the original black and white negative.

finger. A narrow opaque screen for a set LUMINAIRE.

finishing services. See COMPLETION SERVICES.

fire up. Set the equipment into motion.

first answer print. The print that is first viewed in its entirety by the producer, who will judge its acceptability.

first cameraman. The principal camera operator.

first feature. The more important film shown on a double bill in a motion-picture theater, usually the one having more popular stars.

first generation duplicate (dupe). A duplicate made from the original film.

(The second dupe is made from this, the third is a duplicate of the second, etc.)

first grip. The stagehand who is in charge of the other stagehands on a motion-picture production. See GRIP.

first prop man. The person who is in charge of the members of the PROP crew on a film production.

first run. The initial showing of a motion picture by exhibitors.

first trial composite. See FIRST ANSWER PRINT.

fishbowl. The booth in a television studio from which persons can observe the filming, usually reserved for sponsors and other VIPs.

fisheye lens. See BUGEYE.

fishpole, fishing rod. A long microphone boom, manipulated by hand.

525-line. The standard number of horizontal sweeps per frame used in television transmission systems throughout the Western Hemisphere and in Japan.

fix. To secure the developed image on a film by treating it in a chemical bath.

fixed. Referring to **1.** the developed images that have been run through a chemical bath; **2.** a stationary position of a piece of equipment.

fixed focus. The manner in which a lens holds all subjects in focus regardless of the distance setting.

fixed-focus viewfinder. VIEWFINDER that keeps the subject in sharp focus regardless of its distance from the camera position.

fixed position. The designated and unmovable time slot for a television commercial; sold at a premium rate.

fixing. The process of bringing up the developed image on a film by treating it in a chemical bath.

FK. A 5-kilowatt lamp.

flag. A square or rectangular opaque sheet (GOBO) used to block light from the camera lens, from portions of the set, or wherever it is not wanted.

flagging. A television picture distortion caused by inaccurate playback HEAD timing coordination on videotape.

flagship. The principal station of a television network.

flag stem. The stem attached to a small rectangle of wood or cardboard used to shield a camera lens from direct light or to shade a portion of the set.

flapper. See SWINGER.

flare. 1. A dark "scorched" area on a television picture tube created by oversaturation of light. **2.** Film areas that have been accidentally exposed to light through camera components that have not been light-tight.

flash. A bright spot caused by overexposure in the film frame or by some undesirable reflection.

flash ahead. See FLASH FORWARD.

flashback. A shot or scene in a motion picture which depicts an incident or event that occurred before the time shown but not necessarily before the beginning of the picture. (It can represent an actor's "remembrance" of an earlier event in the film).

flash cutting. The use of very short shots in a sequence.

flash forward. A shot or scene that introduces an event that will occur further along in the motion picture.

flash frame. 1. A frame briefly shown that injects a new subject into a shot. **2.** The intentional overexposure of a film frame to provide a visual cue for editing purposes.

flashing. See RE-EXPOSURE.

flash pan. See SWISH PAN.

flat. 1. A wooden or plywood framework on which a painted canvas scene has been stretched to serve as a lightweight movable background for action shots. **2.** Film that lacks video contrast.

flatbed editing machine, flatbed editor. A table equipped with vertical spools for film and sound tracks and other equipment used in the editing process.

flat glass. A glass sheet made with both surfaces so smooth that optical distortions can't occur.

flat light. The lighting of a subject or object so that sharp contrast is avoided.

flick, flicker. Slang for motion picture.

flicker. 1. Variations in light intensity coming from the projector. **2.** A temporary loss of sharp visual images caused by a decrease in projector speed or by shutter malfunction.

flick pan. See SWISH PAN.

flies. The area in a studio, above the set, used for storage or for raised LUMINAIRES.

flighting. Nonconsecutive calendar dates on which broadcast advertising *for a particular product* is to be shown.

flip. 1. To show a new FLIP CARD before the camera. **2.** To rotate the lens mount.

flip cards. 1. CUE CARD. **2.** Cards on which program titles and credits are printed, stacked in a looseleaf binder for rapid flipping before the camera.

flipover. An optical illusion that makes the picture seem to turn from left to right (or vice versa) to show a picture on the reverse side.

flippers. See BARN DOOR.

flip stand. The stand on which the FLIP CARDS are placed before the camera.

flip wipe. See FLIPOVER.

floating release. A motion picture that is available to all theaters rather than one booked for CIRCUIT release only.

flock paper. Opaque paper used for MATTES in the production of special effects.

floodlight, flood. Any LUMINAIRE that lights a wide area without casting shadows.

flopover, flop. An optical illusion that makes the picture seem to turn from top to bottom (or vice versa). See FLIPOVER.

flop sweat. A performer's stage fright, which causes sweating palms and/or brow.

floor. The area in a studio or set where the performance occurs.

floor manager. The person who directs studio activities on the FLOOR and is in contact with the control room by headphones.

floor men. Stagehands or GRIPS.

floor plan. 1. A bird's-eye-view scale rendering of the set. **2.** The arrangement of properties and scenery on the studio set.

flub, fluff. 1. To make an error in the delivery of a line of dialog or commentary. **2.** The error itself.

fluid gate. See WET GATE.

fluorescent bank. Four to six fluorescent tubes mounted in a flat boxlike reflector case.

fluorescent light. Light produced by a gas-filled tubular lamp that has a phosphor coating.

fluting. The distortion or bending of film edges, caused by tight winding and/or excessive humidity.

flutter. 1. Distortion in either the picture or the sound caused accidentally during exposure, printing, or projection. **2.** Distortion in picture or sound caused by irregularities in the tape speed.

flux. The flow of light measured in LUMENS.

fly. 1. To suspend, support, or store scenery from cables or ropes. **2.** The area above the set in which scenery of LUMINAIRES are lifted or stored.

f-number. The number attained by dividing the focal length of the camera lens by the diameter of its APERTURE, a procedure necessary for sharp focusing.

focal length. The distance from the optical center of the camera lens to the film FOCAL PLANE when the lens is focused at infinity.

focal plane. The plane that passes at a right angle or perpendicular to the camera lens' optical axis and on which the image appears when the lens is focused at infinity.

focal plane matte, focal plane mask. The MATTE or mask placed in a camera slot and positioned in front of the film.

focal point. The point behind the camera lens at which an object in the action area comes into clear focus when the lens is set at infinity.

focus. 1. The clarity of an object obtained with a camera lens adjusted to a specific point in the vision field. **2.** To make such a adjustment. **3.** To direct an electron beam or light ray so that it comes to a fine point.

focus band (ring). The part of a lens cylinder that is rotated in order to focus the lens; inscribed with a calibrated scale for distance, shown in feet or meters.

focusing. Bringing the subject into clear view by adjusting the distance from lens and film.

focusing viewfinder. A VIEWFINDER in which the focus is changed continuously as the camera distance in the action area changes.

focus plane. The plane at which a lens will form a clear image when the lens is focused at a point closer than infinity. Cf. FOCAL PLANE.

focus pull. To refocus the lens during filming so that part of an image farther from (or nearer to) the camera is brought into sharp focus, thus allowing the original subject to become soft or slightly off focus.

fog. 1. To create the illusion of fog by artificial means. (See RUMBLE POT.) **2.** The illusion itself. **3.** To ruin undeveloped film by exposure to light.

fog filter. A special diffusing filter on a camera, used to provide a softer photographic effect; also used to create the illusion of fog.

Foley. A process for the synchronous replacement of human or animal sounds incidentally made during the shooting of a scene in order to achieve perfect SYNC between sound and action, named for the man who invented the process.

Foley state. A facility equipped with devices and equipment needed to create sound effects to accompany a shot or scene: e.g., a water tank for splashing sounds; cement sidewalks and dirt runways for the sounds of footsteps respectively on city streets or in the woods.

Foley tracks. The reels on which FOLEY sounds are recorded, later to be mixed and added to the film during the rerecording (dubbing) process.

follow action. See FOLLOW SHOT.

follow focus. Readjustment of the lens focus as the subject of a shot changes place or direction in order to keep the subject in sharp focus. Also necessary when the camera itself is moved.

follow shot. A shot made by the camera as it follows the moving subject.

follow spot. 1. A high-power, narrow-beam LUMINAIRE which is focused at performers as they move from place to place. **2.** To move the camera as it tracks performers in motion.

foot. The end of a film or tape reel.

footage. A length of film measured in feet or meters.

footage counter. An indicator that shows the number of feet of film run through a camera.

foot candle. The number of LUMENS per square foot of light; the measurement of light intensity in a specific area.

footprint. The area covered by a SATELLITE.

force process, force develop. To develop film beyond the recommended period of time in order to reduce overexposure.

foreground. The action area that is nearest the camera in a shot or scene.

foreign release. The distribution of a motion picture in foreign countries.

foreign version. An American-made film dubbed for foreign markets.

forelengthening. An illusion of greater depth in shots made with wide-angle camera lenses.

foreshortening. An illusion of depth reduction in shots made with telephoto lenses.

format. 1. The organization and style of an event to be photographed. **2.** The order in which performers are to appear in a production.

"Fortnightly Decision." A ruling by the United States Supreme Court (1968) that permits operators of closed-circuit television to rebroadcast programs without copyright restrictions. Named for the decision handed down by the Court in a suit brought against the Fortnightly Corporation by United Artists.

foundation light. See BASE LIGHT.

four-walled set. A set (unlike the usual open-sided), in which both camera and performers are enclosed within four walls.

four-walling. An arrangement between an EXHIBITOR and a DISTRIBUTOR in which the latter pays the former a stated price for the use of his theater. All profits above this price belong to the distributor, who sets the price of the tickets and is in charge of all advertising.

foxhole. The small sprocket holes used on positive CinemaScope film prints when copies with four magnetic sound tracks are needed; named for Twentieth Century-Fox Company, which originated the process.

frame. 1. A single shot in a film or the space it occupies in the lens aperture. (A motion-picture camera does not record continuous action but records in separate images [frames], which when projected on the screen give the illusion of motion. Thus a frame is actually a component of a shot.) **2.** To register film in the GATE of a projector through a special mechanism. **3.** To organize a specific shot.

frame counter. A dial on a film counter which shows the number of frames in a single foot of film.

frame frequency. The number (30) of transmitted frames per second.

frame glass. See PLATEN.

frame line. The thin horizontal line that divides two frames on a FILMSTRIP.

frame memory. Receiver circuitry that stores single transmitted frames for continuous retrieval.

frame puller, framer. A projector device used to raise or lower FRAME LINES in order to remove them from view.

frames per minute. The number of frames that are exposed in a camera, or that pass through a printer or projector, in one minute.

frames per second. The number of frames that are exposed in a camera, or that pass through a printer or projector, in one second.

frame store. An analog-to-signal converter that stores thousands of frames or still pictures on magnetic disk packs.

frame-up. 1. To position the camera in order to achieve better photographic balance for the scene to be shot. **2.** To align film accurately in the projector GATE.

framing. The manipulation of camera positions in order to achieve the best composition for a shot or scene to be filmed.

free lance. Any person, such as a producer, screenwriter, designer, or cameraman, who sells his/her professional services independently.

freeze. 1. The final approved version of a format. **2.** DRAW A BLANK.

freeze frame. To continuously print a single frame of film (a) so that its still image can be held on screen, to be studied at leisure, or (b) in order to produce the effect of stopped action or a still photograph when seen on the screen.

frequency. 1. The rate at which air molecules are compressed cyclically by a vibrating object, electronic generator, or air column for the production of sound. **2.** The rate at which an electronic impulse (or light or sound wave) moves past a specific point during a specific period of time. **3.** The wave length of a broadcast transmission. **4.** The average number of times viewers are exposed to a program series or to the same commercial over a specific time period.

frequency discount. A special low rate offered to advertisers who schedule commercials over an entire thirteen-week cycle or at an agreed-upon minimum number of times per week.

frequency equalizer. A device used to enhance audio quality, primarily by the suppression of one of the five frequency ranges.

frequency response. The ability of a recorder or recording equipment to transmit varying frequencies of a signal.

Fresnel, Fresnel lens. 1. A lightweight condenser or PLANO-CONVEX lens used as a condenser for spotlights. **2.** Any spotlight with a Fresnel lens.

friction head. A rotating CAMERA MOUNT which provides smooth rotation motion for the camera.

frilling. The loosening of film emulsions edges from the base.

fringe area. The remotest distance of an area at which broadcast signals from a station are received.

fringe evening time. The television broadcast hours, 5:00–7:00 P.M. and 11:00 P.M. to 1:00 A.M., that precede or follow PRIME TIME.

fringing. The definition loss surrounding an image that has been matted into a background. See MATTE.

from the top. A term used in rehearsals, meaning to repeat the performance or a portion thereof. Occasionally it means literally to begin again at the top of a page of script or music.

front end. Referring to production costs that occur before the actual production begins.

front lighting. Any lighting that comes from the camera position or close to it.

front projection, front-projection process. Any means used to project a background image along the lens axis directly on the actors and on a screen behind them. The image on the actors is then eliminated by the use of a special lighting technique. Often used to achieve special effects, the technique produces images without shadows in which the photography of studio action is combined with a projected background scene, which may be a still photograph or a motion-picture image.

frying pan. A screen used on a set to diminish or soften the light.

f-stop. The lens aperture setting that indicates the amount of light that passes through the lens by dividing the FOCAL LENGTH by the diaphragm diameter. The lower the setting (or number), the greater the exposure. Moving to the next higher number decreases the exposure by half.

f-stop band, f-stop ring. A ring on the lens barrel which turns to change the IRIS setting. F-STOPS are inscribed on this band.

FTC (Federal Trade Commission). The United States government agency that is responsible for the regulation of television advertising.

full aperture. The IRIS on a camera lens opened to maximum circumference.

full coat, full coat magnetic film. See MAGNETIC FILM.

fuller's earth. A highly absorbent clay powder used to simulate dust on a production set.

full net. A television network hookup of all affiliate stations.

full network station. Any AFFILIATE that carries a minimum of 85 percent of the PRIME TIME programming offered by its network.

full shot, full figure shot. 1. A camera shot in which the entire body of the performer is shown. **2.** To include all of a subject, such as a yacht, a barn, or a group of people so that the image entirely fills the frame.

full track. Any recording that has used all the available surface of an audiotape.

fungus spots. Blemishes on a filmed image caused by a fungus growth on the film emulsion.

funnel. See SNOOT.

fuzzy. Description of an out-of-focus image.

G

gaffer. The chief electrician in charge of lighting equipment.

gaffer grip. See BEAR TRAP.

gaffer tape. Adhesive tape sensitive to pressure, used on set or location rigging.

gag. 1. A joke or comic bit used in a motion picture or television production. **2.** A special effect. For example, a car wheel rolling off in a scene is called a wheel-off gag.

gain. Audio amplification.

gain control. An audio amplification control.

galvanometer. An electrical mechanism in which mechanical movements are produced through variations in current.

galvanometer recorder. A sound recorder in which a GALVANOMETER is used to produce variations in the pattern or the intensity of a light beam as it strikes the sound-track area.

game show. A television broadcast show in which contestants, often from the audience, compete for money and/or prizes.

gamma. 1. The maximum contrast gradient of developed film. **2.** The contrast ratio between the input and output of a camera.

gang. Two or more switches on one control.

gangster film. A motion picture based on underworld characters and their exploits.

gang synchronizer. A device used to achieve synchronization among ORIGINALS, WORKPRINTS, and SOUND TRACKS.

gap. A minuscule space between two poles of a MAGNETIC RECORDING,

PLAYBACK, or ERASE HEAD. (Approximately 1/10,000 or 1/20,000 of an inch.)

garbage. The interference of video and/or audio signals on adjacent frequencies.

gas. To pressurize the BOOM tank of a camera.

gate. 1. In a camera or projector, the APERTURE in which the frame is exposed or projected. **2.** The support for the PRESSURE PLATE in a camera or projector.

gator grip. See BEAR TRAP.

gauge. The width measurement of any standard film, usually counted in millimeters (35mm, 16mm, etc.).

gauss. A unit of magnetic induction.

gauze. A thin material used to diffuse strong light or to reduce excessive light on a subject.

gear head. A camera mount with wheels which is used to achieve a smooth motion for PAN and/or TILT shots.

gel(atin). A translucent sheet of colored filter material used to change the color characteristics of a light source.

gelatin filter, gel filter. A sheet of colored gelatin used for light filtration on a camera.

genealogy. The change in the EMULSION POSITION on successive duplicates of the film. (The original film is known as the *first generation.*)

general release, general showing. The exhibition of a motion picture in theaters throughout the country, as opposed to premiere engagements or limited engagements in selected theaters.

generating element. A microphone TRANSDUCER.

generation. A stage in the duplication of film, in which each successive DUPE suffers a loss of quality. See GENEALOGY.

generator. A dynamo operated with diesel fuel or gasoline which generates alternating currents.

generator truck. A vehicle that transports the GENERATOR.

Geneva movement. A mechanism used to hold frames motionless in a camera, projector, or printer.

genlock. A device used to synchronize signal sources.

genre. A motion-picture category, such as Western, mystery, gangster, or sci-fi films.

geometry errors. Aberrant dimensional changes in videotape velocity.

geosynchronous satellite. A SATELLITE that orbits at the same speed as the rotation of the earth.

get in character. A director's instruction to actors to assume the roles they are playing, in preparation for filming.

Gev. One billion electron volts.

ghost, ghost image. 1. A hazy image made by double exposure or double printing to give the illusion of a ghostly presence in the scene. **2.** An accidental "ghosting" of the image during shooting or projection. **3.** A secondary image on the picture tube created by reflected transmission signals.

GHz. See GIGAHERTZ.

gigahertz, GHz. One billion HERTZ.

gigawatt, GW. One billion WATTS.

gimbal mount, gimbal tripod. Any camera support that has a free-swinging mount used to maintain a level position for the camera.

gimmick. 1. Originally, a device used by a carnival pitchman to control his machine in order to delude the player or by a magician in the performance of a trick. Now, any trick or plot device used to rivet the attention of the audience, sometimes employed in the final solution of the conflict.

glass filter. A photographic filter made of a GEL inserted between two sheets of glass.

glass painting. Scenery painted on a sheet of glass.

glass shot. 1. A shot made through a sheet of glass on which artwork or titles are superimposed so that they may appear against the action in the background. **2.** A glass sheet on which a scene is painted and held before the camera so that it will appear to be on the same scale as the distant life-size scene and will merge with it when seen through the clear portion of the glass. Both used in SPECIAL-EFFECTS shots.

glossy. A photographic print having a shiny surface.

gloving sound. See VELVETING SOUND.

glow light. A dim light along the outlines or outer edge of an object.

gobo. 1. An opaque screen used before a set light to block light from any area or surface on which it is not wanted. **2.** A screen or material used to absorb sound.

gobo stand. See CENTURY STAND.

gofer, gopher. A production assistant who runs errands (e.g., "goes for" coffee).

golden time. Sundays, holidays, or other special occasions on which union members are paid *more* than overtime (usually time-and-a-half) wages.

Goldwynism. A verbal gaffe said to have been made by early Hollywood producer Samuel Goldwyn: e.g., "Include me out."

gooseneck. A flexible stand for a microphone.

gothic movie. A costume film based on a gothic novel, one in which the general atmosphere is melancholy and/or eerie, and the locale (e.g., an ancient castle, desolate moors, or medieval village) is steeped in mystery. Examples include *Frankenstein* and *Wuthering Heights.*

go to black. To gradually fade from the visible image to a dark or blank screen.

gradation. The degree of tone change in a photographed image.

grade. To measure the variations in the density of shots and scenes which are caused by differences in time or lighting conditions, and to mark them so that the PRINTER LIGHT is automatically regulated to produce projection prints of more uniform density.

graduated filter. A FILTER in which the colored section fades into a clear section.

grain. The molecular composition of film EMULSION.

graininess. A photographed image that seems to contain minute particles, or has a pebbly surface caused by particles of silver salts in the emulsion.

G-rated. The label given by the Code and Rating Administration rating board of the Motion Picture Association of America to motion pictures suitable for general viewing.

gray base film, gray base stock. Film with a gray base that reduces HALATION.

gray card. A card having a gray MATTE finish on one side and a white finish on the other. The gray side is read with a reflected-light meter to determine the optimum F-STOP; the white side is used when illumination is not sufficient to read the gray side.

gray scale. A cardboard strip containing a scale of shadings ranging from white to black.

grease-glass technique. The process of shooting through Vaseline-smeared glass to produce a blurred effect on part of the scene being filmed.

grease paint. The cosmetics used by actors when making up their faces (or bodies) for the roles they are playing.

green. PROJECTION prints that have not hardened or dried.

greenery. Any foliage such as bushes, small trees, or flowers used on a set or location as part of the scenery. Can be artificial or real.

green print. A POSITIVE PRINT that has not been waxed or run through a projector.

green room. The off-stage room in which performers or interviewees wait until called to appear before the cameras.

greensmen. The crew members who take care of the GREENERY on the set or location.

grid. 1. In a studio a latticework of steel pipes suspended horizontally overhead to which supports for lights and scenery are attached. **2.** A television alignment chart.

grille cloth. A cover for a loudspeaker.

grip. A set handyman who does on-the-spot carpentry, lays DOLLY TRACKS, moves scenery, etc.

grip chain. A small-linked chain used on a set for a variety of purposes, such as hanging scenery.

grip truck. A hand-operated wheeled vehicle used to transport lighting equipment in a studio.

gross rating points (GRP). The total number of rating points for a specific advertising schedule, without taking viewer DUPLICATION into consideration.

ground. The absolute lowest voltage point in an electrical system.

ground glass. The translucent glass sheet on which an image is seen in the viewfinder on a motion-picture camera.

ground glass viewfinder. A VIEWFINDER in which the image is seen on a ground glass sheet instead of through an optical system.

ground row. See COVE.

ground wave. The portion of a broadcast signal that follows the contours of the ground.

guard band. The section of audiotape or videotape that separates the different recording tracks.

guide track. See SCRATCH TRACK.

guillotine splicer. A perforating and trimming device used in film editing, by which a roll of tape extremely sensitive to pressure is applied across the film to align two strips for splicing.

gun. A CATHODE RAY TUBE source that emits a narrow beam of electrons that can be sharply focused.

gyro head. A TRIPOD mount that minimizes any sudden camera movement.

H

hair stylist. The person who arranges the hair of actors before a performance.

halation. 1. A hazy image or halo that is seen around an object or subject in a film, caused by reflected light extending beyond the desired boundaries. **2.** A print flare caused by excessive light reflected from the FILM BASE through the EMULSION.

half-apple. A sturdy, low wooden box used as a means to achieve a desired position for an actor or object in a camera shot. Also used to increase the height of an actor or object. Approximately half the height of an APPLE BOX.

half broad. A 1,000-watt floodlight shaped like a box.

half track. 1. An audio recording that has used only about 40 percent of the available surface of tape. See FULL TRACK. **2.** The optical or magnetic SOUND TRACKS combined in a single print.

halide. See SILVER HALIDE.

hand camera. A motion-picture camera small enough to be operated by hand without use of a DOLLY, TRIPOD, or other mount.

hand-drawn traveling matte. A MATTE strip made by photographing drawings of silhouettes taken from action frames and superimposed on other action, finally to be BIPACK printed.

hand-held. 1. Referring to a shot made with a hand camera, **2.** the imperfect image that occasionally results from such a shot.

handlebar mount. A CAMERA MOUNT that has two handles so that the cameraman can use both hands when manipulating the camera.

hand model. A performer whose hands only are used in a shot.
handout. A publicity release.
hand properties, hand props. Small objects handled by actors in a shot or scene.
hands-on animation. A process in which drawings done frame-by-frame by hand are photographed on film and then stripped into the live action; e.g., a swarm of real bees attacks a house, but only the animated replicas "attack" the performers inside the house. The animated figures are drawn by hand before being traced on CELS.
happy talk. Incidental chit-chat on a television news program.
hard. Strong contrast appearing in shots.
hard copy. A paper CATHODE RAY TUBE printout.
harden. To sharpen the focus of the lens.
hardener. A chemical bath element that hardens the GELATIN holding the EMULSION.
hard-front camera. A camera with a single hole, instead of a movable TURRET, in which the lens is attached.
hard light. A narrow spotlight beam used to produce sharp shadow edges.
Hard Rock. Nickname for the New York headquarters of the American Broadcasting Company.
hard-ticket roadshow attraction. A feature film presentation to which reserve seats are sold.
hardware. 1. Equipment used on a set or location. **2.** Electronic equipment.
harmonic. Describing a signal whose FREQUENCY is an integral multiple of its original related frequency.
having had. Direction (or suggestion) given to cast and crew in a reporting call so that they can determine their meal schedule; e.g., "Report at 11:00 A.M., having had" means that there will be no lunch break for hours, so they should eat not long before 11:00 A.M.
haze filter, haze-cutting filter. A lens filter that removes ultraviolet light and diminishes any haziness.
HBO (Home Box Office). A television cable service owned by Time Inc.
head. 1. The starting end of a reel of film or tape. **2.** The GATE on an editing machine.
head end. The antenna point of origin for cable television transmission.
head gaffer. The chief electrician on a film production.
head grip. See FIRST GRIP.
head leader. Any one of the several kinds of LEADERS used at the start of original films or prints.
headlife. The average life of a video recording head from one adjustment overhaul to the next.
head-on shot. An action shot in which the actors, vehicles, horses, etc., are moving directly toward the camera.
head out. Describing film that is wound so that the first frame of the sequence is on the outside of the roll.

headphones. Tiny wired receivers held over each ear of the RECORDIST by a flexible clamp; used to judge the quality of sound.

headroom. The area between the top of an actor's head (or the top of an object) and the upper edge of the FRAME.

headset. A small wired communication system composed of an earpiece and mouthpiece mounted on a head band, worn by studio technicians.

head sheet. Photographs of an actor taken from different angles, on a single STILL print.

head shot. A shot in which only the actor's head (and occasionally shoulders) is framed.

heads out. Describing a reel of tape or film that is ready for projection.

heat-absorbing filter, heat filter. A filter that can absorb infrared radiation; used in projectors to reduce the amount of heat on the film.

heater barney. An electrically heated padded cover for a camera, used to protect it during filming on location in cold weather.

heavy. A villain in a film.

hectohertz. 100 HERTZ.

height. The vertical dimension of a television picture.

helican scan. A signal recorded horizontally on videotape, easy to edit by splicing along horizontal lines.

helicopter mount. See COPTER MOUNT.

helios noise. The five-minute interference in wave transmission that occurs when a SATELLITE passes between the tracking earth station and the sun.

henry. A measurement unit of electrical energy storage in a magnetic field.

hero, heroine. The principal performer in a motion-picture or television film around whom the conflict is built and by whom it usually is resolved.

hertz. The frequency unit of sound equal to one cycle per second (cps), most often written as Hz.

HFR (hold for release). Instruction to hold any material from broadcasting until further notice.

hiatus. A deliberate interruption in a series of scheduled commercials in order to stretch an advertising budget.

HID (high intensity discharger). Mercury, metal halide (AC current only), and high-pressure sodium lamps.

hi-fi, high fidelity. Uniform frequency response in sound reproduction.

high-angle shot. A shot made from a camera placed above the action. The camera may be only a few feet above the subject for a close-up, or hundreds of feet above it for a long or ESTABLISHING SHOT.

high contrast. 1. A sharp change from white to black values in an image. **2.** A type of film used in printing to eliminate the EMULSION background from title MATTES.

high grain. The signal level of one volt or more.

high hat, hi hat. A short tripod used for low-angle camera shots.

high-intensity arc, high-intensity carbon arc. A LUMINAIRE that produces an extremely bright light beam through the use of a transformer and a carbon arc.

high key. Referring to bright illumination in which the lighter (or gray) tones are predominant, giving a bright effect. Generally used for comedies.

high-key lighting. Any lighting used to achieve HIGH-KEY images.

high level. See HIGH GRAIN.

highlight. The maximum possible illumination of an image without loss of detail.

highlight density. The density of the lightest area in a positive filmed image (or the darkest area of a negative image).

high-pass filter. An audio circuit electronic filter used to attenuate all frequencies that fall below a desired frequency.

highs, high frequency. Sound frequencies in the approximate range of 15,000 HERTZ.

high shot. See HIGH-ANGLE SHOT.

high-speed camera. A specially designed camera that operates at greater than average speed, used to achieve slow-motion effects when the film itself is run at normal speed.

high-speed cinematography, high speed. Any shooting that is done at greater than normal speed in order to produce extreme slow-motion film when it is projected. Also used in examination of the frames for future editing.

high-speed duplication. Rerecording a copy from a master tape at a speed much greater than the original.

high-speed film. 1. Film that has extra perforations for use in a high-speed camera. **2.** Film that is extremely sensitive to light. **3.** Film made with a high-speed camera.

high-speed photography. See SLOW MOTION.

hiss. Noise on tape playbacks, that is caused by a distortion in high-frequency signals.

historical epic. A full-length film in which the plot is built around historical events and persons. See EPIC.

hit. A brief audio interference.

hitchhike. A commercial message shown after a program ends.

hit 'em all, hit the juice. A request to turn on all lights to be used during filming.

hit the mark. Direction for a performer's movement to a predetermined spot on the set where lights and cameras have been arranged for shooting. Sometimes there is a chalk mark to designate the exact spot.

HMI lamp. A specially designed lamp used to produce light similar to bright sunlight.

hold. 1. To make reprints of a single action frame. **2.** Any animation CEL that is not changed from frame to frame. **3.** A taped performance that is to be held for possible use.

Hollywood blacklist. A list of professionals (writers, directors, actors) in the motion-picture industry who were refused employment for some time after the House Committee on Un-American Activities (1947) accused them of being members or fellow travelers of the Communist Party.

Hollywood Ten. A group of individuals in the motion-picture industry who refused to testify before the House Committee on Un-American Activities (1947) and in consequence were sentenced to prison for one year and blacklisted for some time thereafter. They were Alvah Bessie, Herbert Biberman, Lester Cole, Edward Dmytryk, Ring Lardner, Jr., John Howard Lawson, Albert Maltz, Samuel Ornitz, Adrian Scott, and Dalton Trumbo.

hologram. A laser-produced photographic image that appears three-dimensional.

holography. LASER photography.

holy factor. A high-key illumination used in color photography.

Home Box Office. See HBO.

homes. Households in which there are one or more TV sets, used in audience survey counts.

home video recorder. Receiving and recording equipment used to record television programs for later playback. Owners and/or producers of the programs claim the use of HVRs to be illegal and an infringement of the copyright laws under which their products are protected. The matter is pending in the courts.

hook. A provocative incident or piece of dialog used to capture the instant interest of the audience in the opening shots of a film.

hooking. The distortion of a television picture due to uncoordinated timing in the tape or playback HEAD.

horizontal. A television scan line signal. See SCANNING LINE.

horizontal blanking. The elimination of signal transmission during horizontal retrace.

horizontal resolution. The ability of a camera to denote changes of intensity along a single scan line. See SCAN LINE.

horizontal saturation. A heavy advertising schedule that uses the same time period daily for a certain length of time.

horror film. A category of films that emphasizes the macabre and is intended to provoke reactions of terror from the audience.

horse opera. A Western film.

hot. 1. Referring to an image that is too brightly lighted; **2.** a person (actor, director, writer) whose popularity is at a peak; **3.** a property, such as a best-selling novel, which filmmakers vie to control or purchase.

hot box. A box into which lighting cables are plugged.

hot frame. A frame that has been intentionally overexposed to provide a visual cue in editing.

hot points. The sharp edges of a camera tripod.

hot-press titles. A technique used for making titles in which heated type is pressed against white, black, or colored transfer paper that is then pressed against a title board or CEL.

hot splice. A film splice made on a HOT SPLICER.

hot splicer. A splicing device in which the metal is electrically heated in order to warm the spliced ends and speed up the action of the film cement.

hot spot. An area of excessive reflection from a portion of an illuminated object.

house agency. An advertising agency that yields control of certain functions, such as choosing the director of a TV performance, to the sponsoring client organization.

housecleaning. General dismissal of personnel and hiring of replacements.

household. Any home in the United States that has one or more television sets.

housewife. For purposes of television-audience surveys, any female who runs a household.

housewife time. See DAYTIME.

How many pages? Reference to the film script pages to be shot that day ("How many pages today?").

hub. A rotating cylinder used to rewind tape or film onto a core.

hue. 1. Color and its gradations. **2.** A distinctive color wave length (black, white, and gray do not have hue).

hum. An intrusive low-frequency sound caused by inaccurate circuit alignment.

human interest. Any subject matter on television or in films that has wide appeal to an audience; something that evokes empathy.

HUT (homes using television). An audience survey count designating the number of broadcast homes in which television is viewed during an average fifteen-minute time period.

hype. To promote, generally with excessive enthusiasm, a motion picture in a variety of ways, including personal appearances by the stars, wining and dining critics and distributors, full-page newspaper ads.

hyperfocal distance. The distance between a camera lens set at infinity and the closest object held in acceptable focus.

hyphen. A person who functions in two or more capacities on a film production: e.g., writer-director or actor-writer-director.

hypo. 1. See SODIUM THIOSULFATE. **2.** To schedule popular programs during the rating periods for television productions.

Hz. See HERTZ.

I

IATSE (International Alliance of Theatrical State Employees). The trade union of persons who work behind the scenes on film sets.

IBEW (International Brotherhood of Electrical Workers). A trade union of electrical technicians.

iconoscope tube. A television camera tube in which there is a minimum of IMAGE LAG when operated at an average-light level; invented and patented by Vladimir Kosma Zworykin. See IMAGE ORTHICON TUBE.

ID. See STATION IDENTIFICATION.

identification. The EMPATHY an audience feels for one or more characters in a film.

idiot card. See CUE CARD.

if (intermediate frequency). The standard frequency for the electronic signal waves in the receiver.

Ikegami. A Japanese electronics company that manufactures hand-held motion-picture cameras.

illumination. Any source of natural light (such as the sun) or artificial light used in filming.

image. The picture that appears on the screen or is seen through the camera.

image degradation. The blurring of detail or loss of sharp contrast in a photographed image.

image distortion. Unwanted change in an image due to some fault in the camera lens or in the focus.

image duplication. The showing of several identical images on separate parts of a frame by use of an optical attachment on the camera or projector.

image enhancer. A signal processor that achieves a sharper image by bringing into view luminance detail that has been obscured.

image intensifier. An electronic lens attachment used between the lens and the camera to improve low light levels.

image lag. An image that remains on the screen a few seconds after the camera has been moved.

image orthicon tube. The standard television picture tube, invented by Vladimir Kosma Zworykin, called the "father of television"; he also invented the ICONOSCOPE TUBE.

image pickup tube. A device that converts optical images into electrical signals through an electronic scanning operation.

image plane. The plane at which an image is clearly seen in the lens when it is focused at infinity.

image point. A point behind the lens at which an object in the action area comes into clear focus when the lens is set at infinity.

image replacement. A special-effects process in which parts of an image in a shot are removed and replaced with other images.

Imax. A film production technique in which the screened picture is 70 feet high and 135 feet wide, surrounding the audience; it employs six separate sound tracks, nine speakers, and 20,000 watts of audio power. In 1982 sixteen Imax films were showing throughout the world at the twenty theaters equipped to show them.

imbibition. The last dye-transfer stage in the processing of Technicolor prints.

impedance. The resistance to the flow of an AC (alternating current) in an electric circuit.

impedance matching. To ascertain that any connection or attachments that are secured to an audio circuit are identical to the IMPEDANCE in the circuit.

impedance-matching line transformer. A transformer used in a cable to achieve impedance matching.

impedance-matching transformer. A TRANSFORMER that is attached to audio circuits to circumvent IMPEDANCE MISMATCHING.

impedance mismatching. The attachment to an audio circuit of a component whose IMPEDANCE does not match that of the audio circuit.

imported signal. See DISTANT SIGNAL.

impressionism. The creation of a general impression by joining a series of shots of subjects which are unrelated in time or space or both.

impressions. The gross program or commercial audience.

improvise. To spontaneously create dialog or action that has not been written into the script.

in. To move toward. (The actor moves toward the camera or vice versa.)

in-betweener. An artist who draws the figures or scenes between the major images created by an animator.

in-betweens. Animated drawings made by the IN-BETWEENER.

in-camera matte shot. A camera shot in which a portion of the action area is blanked out with a stationary black mask placed in front of the camera in order to photograph other action on the same film, using a second mask.

incandescent lamp. 1. A gas-filled lamp in which light is produced through a tungsten wire filament in a sealed glass bulb. (When electrically heated the filament produces a light of intense brightness.) **2.** A LUMINAIRE that uses incandescent lamps.

in character. Referring to **1.** the assumption of a role by an actor as he/she prepares for shooting; **2.** dialog or action that is essential to the characterization.

inching. Moving film through the projector or editing machine frame by frame.

inching knob. A control on an editing machine or projector that is used to move the film forward or backward by frames.

incident. Describing light that falls on a subject from any source.

incident light meter. A device that measures the amount of light that strikes a subject.

inclining prism. A rotating VIEWFINDER.

incoming shot. A shot that is to follow the one being viewed or is the next to enter the editing equipment.

independent, indie, indy. Referring to a broadcast station that is independently operated and offers fewer than ten hours of network programs a week.

independent filmmaking. The filming of productions that are conceived by a person or group not under contract to a major studio; usually done in rented or leased facilities. The film may be produced without union personnel or may be subcontracted to union personnel.

indigenous sound. Any sound that originates at a source that can be seen in the action area of the film.

inductance. The electric energy storage in a magnetic field, generated by the flow of current in a conductor.

induction. The transmission of magnetic or electric currents without direct connection.

industrial film. A film designed to present selected information about a large industrial enterprise, such as automobile manufacturing or coal mining.

infinity. 1. A distance between the camera lens and the subject that is so great the light rays reflected from the subject may be seen as parallel. **2.** A distance setting on a camera lens beyond which all images are in focus.

informational film. A film designed to present facts or knowledge about a specific subject, usually intended for a particular audience.
in frame. See IN SHOT.
infrared, infrared light. 1. The portion of the invisible radiant spectrum that lies beyond the red edge of the visible spectrum. **2.** Film emulsion that is sensitive to light waves that are longer than visible red.
infrared cinematography. Photography that uses film sensitive to INFRARED LIGHT.
infrared film. Any film that is sensitive to infrared radiation. (Infrared film can be used to photograph objects in a darkened room or to film objects through fog or clouds.)
infrared matte process. A MATTE produced by filming a subject before a background that reflects INFRARED LIGHT that is filtered out.
infrasonic. Describing a sound wave that has a frequency below audible range.
ingenue. A young actress or a role for an actress who portrays a young and usually vulnerable character.
in-house. Referring to something that is particular to a specific business: e.g., in-house feuds, in-house jokes, in-house gossip in the motion-picture industry.
in-house unit. A production unit that is under the jurisdiction of the company for which it is filming.
ink-and-paint. The final stage in ANIMATION work following the rough rendering that has been photographed for initial editing.
inker. An artist who applies acetate ink to the animation CEL surface in order to bring out details.
inking. Drawing in the lines needed in the process of producing animation artwork.
inky dink. 1. Any light device that uses incandescent lamps instead of carbon arcs. **2.** A tiny 100–250 watt FRESNEL spotlight.
in phase. The absolute coordination between the movement of the film through the shutter gate and the rotation of the camera shutter.
input. 1. Equipment receptacle to which electrical attachments can be connected so that signals or currents can be fed to the equipment. **2.** Any incoming power or signal.
insert. 1. A shot of a detail that can be made out of sequence and inserted into the principal action during the editing process. **2.** A term used in film scripts to indicate that a particular detail (not part of the shot or scene) is to be inserted in it; e.g., during a battle scene, a map may be inserted for a few seconds to show the area in which the action is taking place. **3.** A segment that is matted into a larger television picture. See MATTING. **4.** Tape or film that is inserted in material made previously.
insert camera. A small camera used for the superimposition of artwork or titles.

insertion. The showing of an individual commercial on a regular advertising schedule.

insertion loss. The loss of signal strength when a piece of equipment is inserted into the circuit.

insert stage. A small studio used for close-up photography of inanimate objects.

in shot. Referring to anything that accidentally appears on the film.

instantlies. The television equivalent of DAILIES.

instant replay. The immediate playback of some portion of a live telecast, occasionally in slow motion or FREEZE-FRAME action. Usually done during sports events such as a tennis match or football game.

instructional film. A film designed to teach techniques, provide instruction, or disseminate information.

insurance. A wide-angle camera shot used to cover up any interrupted movement by the subjects during close-ups; or to prevent the showing of lip movements OUT-OF-SYNC.

in sync. A term meaning that the sound and the action are coordinated.

int., interior. 1. An indoor set. **2.** Designation in a film script to indicate that the action is to occur indoors.

integral reflex viewfinder. A VIEWFINDER built into an inner part of the camera.

integration. The insertion of commercials into a recorded television program.

INTELSAT (International Telecommunications Satellite Organization). The global communications SATELLITE system that includes more than ninety-four shareholding nations with five satellites and 133 earth stations (1978).

intensification. The chemical treatment used to heighten detail in a filmed image.

intensity. The brightness of a light wave or loudness of a sound wave, usually measured by their amplitudes.

interactive television. A system that permits home television viewers, using special push-button equipment, to respond to questions asked by commentators on their television screens. Answers are recorded by computers every six seconds. Available in Columbus, Ohio; soon to be installed in other selected cities.

interchangeable lenses. Lenses with standard mounts that fit any camera with a compatible mount.

intercom. A two-way communication sound system.

intercut. The interfacing of two scenes that show what is happening simultaneously between two persons in different locations; e.g., two persons are talking on the phone and through intercuts the audience is shown the actions or reactions of each of them.

intercutting. The insertion of shots into a series of shots.

interface. The connection made between two or more pieces of compatible equipment.

interference. The disruption of signal transmission by foreign electronic impulses.

interior framing. The use of objects placed in the foreground to increase the dominance of the subject of the shot.

interior lighting. 1. Artificial light used for indoor sets. **2.** The technique used in achieving such lighting.

interior monolog. A technique used to present the unspoken thoughts of an actor in which his/her voice is heard but no lip movement is seen.

interlock. Any system that provides for the synchronization of separated sound track and picture, usually by means of a linked viewer and sound producer.

interlock projector. A projector that reproduces the picture while the synchronized sound is played back on another mechanism.

intermediate. Film used only to make duplicates. A POSITIVE film is made from a NEGATIVE in order to make a second negative; a negative is made from a positive in order to make a second positive.

intermediate reversal negative. A NEGATIVE made from another negative using the REVERSAL process in order to eliminate one GENERATION of printing.

intermittent. A device used to advance and stop frames in the camera, projector, or printer.

intermittent shutter. A rotating lens device used in place of the usual camera shutter.

intermodulation distortion. A distortion of sound that occurs between two or more frequencies when the input and output signals are not compatible.

internal delay. The time needed for a signal to pass through the transmission equipment.

internal rackover. The mirrors in a camera that are used to catch and reflect light from the camera lens into an EYEPIECE LENS.

International Association of Independent Producers. An organization of independent film producers and firms that facilitates the exchange of information, equipment, and personnel among its members; it sponsors seminars and workshops in the United States, Europe, Asia, and Latin America.

International Sound Technicians, Cinetechnicians, and Television Engineers. The trade union that represents the membership of engineers and technicians.

internegative. An optical color NEGATIVE duplicate made from an original POSITIVE. Used for RELEASE PRINTS in order to protect the original film.

interpositive. A POSITIVE duplicate film made from an original NEGATIVE.

Intersync. Equipment for videotape recording used to synchronize the signals of the recorder and the film camera.

intertitles. Any titles shown during a film which are necessary to provide additional information to the audience (e.g., subtitles).

intervalometer. An automatic tripping device used on a camera shutter, adjusted to operate at various intervals of time.

in-the-camera effects, in-camera effects. Interesting VISUALS obtained through processes inside the camera involving MATTES, speed variations, double or multiple exposures, or upside-down shooting. Often used in sci-fi or fantasy films such as *Star Wars* and *Superman.*

in the can. 1. Term used to indicate that shooting on the film has been completed and that the exposed film footage is ready for processing. **2.** Referring to broadcast material that has been recorded and is ready to be aired.

inverter. A device that converts DC to AC current.

invisible cut. A cut made while an actor is in motion, usually by using two cameras or overlapping the action, and later edited so the action appears to be continuous.

invisible splices. Splices made without visible seams between two sections of film.

ips, inches per second. The measurement of the speed at which tape travels.

iris. 1. An old technique once used to achieve a type of dissolve in which a new scene appears in the center of the previous scene and is gradually expanded to fill the screen. **2.** See DIAPHRAGM.

iris (in or out). A WIPE effect generated by a wipe line moving in a circle. For *iris in,* a round spot at the center of the dark screen widens until the picture fills the entire screen (In). *Iris out* reverses the process. Used in silent films but seldom used today except to achieve a particular effect. See IRIS.

irradiation. The scattering of light by the silver grains contained in the film EMULSION.

island. The group of film and slide projectors from which a camera chain is fed.

island position. A television commercial that is separated from other commercials by the program material.

J

jack, jack plug. A plug-in connection used for joining audio components.

jack tube. Telescopic support for a LUMINAIRE that is braced between two walls.

jam. A pile-up of film in a camera due to some mechanical malfunction, causing the operation to stop.

jenny. A power generator.

jingle. A musical commercial.

jog. The frame-by-frame movement of videotape during the HELICAL SCAN editing process.

join. See SPLICE.

joy stick. A hand-operated device that controls the remote operation of electric equipment.

joy-stick zoom control. A ZOOM LENS two-way switch that activates a motor that makes the lens ZOOM in or out.

juice. Electric current.

juicer. An electrician.

jump cut. A sharp advance in an action shot or between two shots when a portion of film is removed, either unintentionally (due to faulty editing) or intentionally to achieve a desired effect, usually to advance the action quickly.

jumper. An extension for a power cable.

jump out. To remove unwanted frames within a scene without loss of smooth continuity.

junction box. A portable TERMINAL used for power cables.

junior, junior spot. A spotlight that uses a 1,000–2,000-watt lamp.

junk. SATELLITES that no longer function but that are still in orbit.

junket. A trip financed by a network or production company for members of the communication media in exchange for publicity.

justified camera movement. Any movement of the camera made for a specific purpose.

justified dolly shot. A camera shot made from a DOLLY, considered necessary to produce a desired effect.

K

Kalvar print. A heat-process POSITIVE film print made from a NEGATIVE on a Metro-Kalvar printer.

keep takes. A term used to indicate TAKES that are to be retained, probably to be used in the completed film.

key grip. See FIRST GRIP.

key in. To electronically MATTE an image.

key light. The principal light or LUMINAIRE that creates the brightest light modeling the subject of a shot.

key numbers. Manufacturers' numbers placed on the edge of film as an aid to processing. They appear every twenty frames (6 inches) on 16mm film and every sixteen frames (12 inches) on 35mm film.

keystone. An image distortion caused when there is an inaccurate angle between the projector and the screen.

kHz. See KILOHERTZ.

kicker, kicker light. A LUMINAIRE positioned so that its light illuminates the back and side of an object.

kidvid. The generic term for television programs for children.

kill. To extinguish lights or eliminate sound.

kilocycle. See KILOHERTZ.

kilohertz. The frequency unit that is equal to 1,000 cycles per second. (Can be written kHz.)

kilowatt, K. 1,000 watts.

Kinescope, Kinescope recording. See TELERECORDING.

Kinetograph. An early motion-picture filmstrip machine developed for Thomas A. Edison by W.K.L. Dickson.

Kinetoscope. A peep-show machine in which moving photographs were shown on a loop of perforated celluloid and could be viewed by only one person at a time. Developed by W.K.L. Dickson for Thomas A. Edison, it was the origin of today's CINEMATOGRAPHY.

Kleig, Kleig light. Early trade name for a variety of lights used in theaters and film studios; now indicates any lighting device used in these places.

knuckle. An adjustable grooved clamping disk on a CENTURY STAND, in which PIPE BOOMS and other equipment can be attached.

L

lab (laboratory). A facility in which exposed cinematography film is developed and printed, sound tracks are processed, and film duplicates are produced.

labels. Words shown in a film to make clear to the audience some information that may not be clarified in the visible content of the film; e.g., a dateline superimposed on a street scene: Paris 1848.

laboratory effects. Special effects that can be obtained through highly technical procedures during film processing.

lace. To thread film into a projector.

lacquer. A protective coating applied to film.

ladies' costumer. A production crew member who is responsible for obtaining and caring for the women's costumes.

laid-in music track. Background music or sound that is natural, not edited into the film after shooting.

lamp. Any device that provides artificial illumination.

lamphouse. A projector component that holds the lamp.

lamp lumens. The maximum amount of light that can be produced by any lamp.

land line. A cable transmission.

lantern slide. See SLIDE.

lap dissolve. See DISSOLVE.

lapel microphone, lapel mike. A tiny microphone that can be clipped to a performer's clothing.

lap switch. A brief DISSOLVE between two video signals.

laser (*l*ight *a*mplification by *s*timulated *e*mission of *r*adiation). A device that generates a long narrow beam of visible electromagnetic waves from 80 to 1,000 TERAHERTZ.

laser videodisc. An audio-visual disc used on special equipment by home viewers that can be "read" and "reread" just as books are. Manipulated by push buttons on a control panel, the disc can be stopped at any point, made to speed up or slow down, and run forward or backward for leisurely perusal or study.

latensification. A process of treating the LATENT IMAGE in order to increase the exposure in the clouded area.

latent, latent image. An exposed but undeveloped film image caused by the exposure of the photographic EMULSION to light passing through the lens.

lateral orientation. A reversed image.

laugh track. Recorded laughter often played during or following lines or routines in a comedy or a comedian's performance.

lavalier, lavalier microphone. A small microphone suspended on a cord around a performer's neck.

lay, lay in. To SPLICE the cut ends of a SOUND TRACK in the editing process.

lay an egg. To give a poor performance.

laying tracks. Positioning sound tracks for the mixing process.

layout. Outline or guide made by an ANIMATOR to be used in plot CONTINUITY.

lead. The principal role in a film.

leader. 1. A length of blank film or tape at the head or tail portion of a roll. **2.** Blank opaque film spliced between sections of a WORKPRINT. **3.** Any blank film needed in identification, editing, or threading processes.

lead-in. A television program that precedes another.

lead-out. A television program that follows another.

lead sheet. 1. A musical accompaniment score. **2.** A horizontal bar graph that registers the precise and intricate relationship between animation action and the accompanying musical beats and/or voice syllables.

LED (light-emitting diode). A glowing crystal chip SEMICONDUCTOR used in television screens. (150,000 LEDs can create a television screen the size of an average wall in a home.)

lens. A transparent glass optical system through which light is refracted on curved surfaces to produce photographic images.

lens adapter. A device used to interchange lenses in a camera.

lens aperture. See F-STOP.

lens barrel. A cylindrical support for a LENS MOUNT.

lens cap. A plastic or metal protective cover placed over the end of a lens.

lens coating. A fluoride coating used to reduce reflection and emit more light through the camera lens.

lens element. Any inner component of a lens.

lenser. Nickname for a CINEMATOGRAPHER or photographer.

lens extender. A device used to position the lens away from the camera for a close-up shot.

lens filter. A colored filter placed in front of a camera lens to change the contrast or tonal value of a scene.

lens flare. A bright spot on a film image that results from the lens receiving too much direct light, either from the sun or from an artificial light source.

lens hood. A camera attachment used to shield the lens from unwanted light sources.

lens marking. The manufacturer's CALIBRATION numbers and index marks placed on a lens to indicate FOCUS, F-STOP, and the DEPTH OF FIELD.

lens mount. Any device used to attach a lens to a camera.

lens prism. An attachment that provides for multiple-image photography.

lens speed. The light transmission capability of a particular lens.

lens spotlight. A LUMINAIRE in which a single sliding lens is used to control the beam.

lens stop. See F-STOP.

lens support. A metal brace attached to a camera mount on one end and near the front of a telephoto lens on the other.

lens turret. A revolving camera mount holding two or more lenses, used to change from one lens to another without delay.

lettering safety. The area within the inner edges of the television set frame in which the titles can be clearly seen.

level. An indication of the amplitude or intensity of sound.

level distortion. Unwanted variations in color saturation and intensity in a television image.

library, library shot. See STOCK SHOT.

library music, library sound. Recorded music and sound effects from films for which they have been especially made; available for reuse, usually on disks.

license. 1. Permission granted by the Federal Communications Commission to operate a broadcast facility. **2.** A business permit that must be secured by film renters from the city tax and permit divisions in each state in order to exhibit films to the public.

lift. To use material from a previously broadcast program.

lifter. A device that lifts the tape away from close HEAD contact during fast forward and rewind operations.

light. 1. Electromagnetic radiant energy made visible. **2.** Any illumination used in shooting a film. **3.** To program the lighting needed for a particular scene or shot.

light-balancing filter. A lightly colored camera or lens filter used to compensate for variations in the color temperature of light caused by changes in atmospheric conditions or in the time of day (e.g., from the heat of noon to the cool of late afternoon).

light board. See LIGHTING CONTROL CONSOLE.

light box. A frosted glass-covered box embedded in an editing table from which a light is projected on filmstrip reels that are run above it.

light bridge. The walkway that runs above the GRID.

light changes. Intensity changes in the PRINTER LIGHT used to correct film that has been underexposed or overexposed.

light 'em out. An order to counteract unwanted microphone shadows by the strategic placement of lights or LUMINAIRES.

lighting. **1.** Any controlled illumination used in lighting the ACTION FIELD. **2.** The natural or artificial source of such illumination.

lighting balance. The correct distribution of light throughout the set.

lighting batten. See BATTEN.

lighting contrast. Artistic variations in the intensity of light as it falls on subject and objects in a shot.

lighting control console. A console equipped with electrical switches from which various lights may be controlled.

lighting grid. See GRID.

lighting plot. A diagram of the positions of lights of various sizes and capacities, made in preparation for shooting.

lighting ratio. The intensity spread between KEY LIGHTS and FILL LIGHTS.

lighting setup. The placement of lights in position before the filming begins on a particular shot or series of shots.

light level. The intensity of light measured in CANDELAS.

light meter. A device used to measure illumination intensity in CANDELAS.

light piping. The blurring of film caused by part of the EMULSION being exposed to light entering from the FILM BASE.

light source filter. A special LUMINAIRE filter used to change its color or color temperature.

light-struck. Film footage that has been ruined by accidental exposure to light.

light-struck leader. Exposed film used as a leader.

light-tight. Referring to any photographic equipment that has been constructed so that light cannot enter it.

light trap. A system of doors (revolving or two-door construction) through which a person can enter a dark room without admitting light.

light valve. A mechanism used in the modulation of light for recording sound tracks.

lily. The standard color chart used to achieve perfect control in film printing.

limbo. A photographic background in which no detail appears and which appears to stretch into infinity.

limited animation. **1.** ANIMATION in which only a portion (e.g., the eyes or mouth) of a figure is animated. **2.** Frame-by-frame photography in which the subject is slightly altered, as is the position of the camera.

limiter (sound). An amplifier device that prevents signals from exceeding a specified limit.

limiter (visual). A circuit that prevents the maximum value of a sound wave motion of a television signal from rising above a specific level.

limpet mount. A camera mount that can be attached to a smooth base by suction cups.

line. 1. One or more words, seldom more than a sentence, spoken by a performer. **2.** The individual beam sweep across a camera image on the picture tube. **3.** Any material that is being transmitted for broadcast.

line amplifier. An AMPLIFIER used to feed a line into the transmission circuit.

linebeat. See MESHBEAT.

line check. A copy made of line material before actual transmission.

line cord. Any wires used in supplying electrical energy.

line feed. Any remote signal transmitted by cable.

line frequency. The number of horizontal frame SCANS transmitted per second, usually more than 15,000.

line monitor. A device on which material being broadcast can be viewed in the control room.

lines. Dialog or monolog spoken by a performer.

line up. To correlate the various elements in filming, such as the camera signals and sound track.

lineup. The broadcast stations that have scheduled material from a single network.

line-up tone. A SOUND TRACK segment that has a consistent signal of one frequency, used to adjust the volume level during rerecording.

lining up, lining up a shot. Placing a camera or cameras so that the ACTION FIELD will be properly covered prior to shooting.

lip sync, lip synchronization. 1. The recording of sound during a performer's speech or song in order to match sound and the lip movements. **2.** A performer mouthing words in synchronization with dialog or songs that have been recorded previously.

liquid gate, liquid immersion gate. See WET GATE.

liquid gate printing. Any printing that is done on a printer equipped with a LIQUID GATE.

literary agency. An agency that represents writers in sales and in contract negotiations.

live-action photography. Photographing living beings in motion as opposed to ANIMATION photography.

live announcer. Usually a local announcer who precedes or follows a recorded broadcast commercial.

live broadcast. A broadcast of something, such as a sports or news event, that is occurring at the time it is being filmed.

live fade. The moving away from the mike of a performer who is still speaking.

live-on-tape. A program that is recorded without pause during its original presentation and is not edited later; it must be filmed so that it fits within a specific time frame.

live recording. An original recording.

live sound. 1. Original sound as opposed to recorded sound. **2.** INDIGENOUS SOUND.

livestock man. A production crew member who cares for the animals that are appearing in a film.

live tag. Commercial information provided by a local outlet, aired following a recorded broadcast commercial related to the product(s) being advertised.

load. To place unexposed film in a camera.

loader. See FILM LOADER.

loader boy. The crew member who is responsible for loading the cameras.

load in. To produce on set all materials needed for immediate filming.

local advertiser. A commercial client who buys time on television, usually to advertise a single product for local retail sale.

locale. The setting, both exteriors and interiors of a film shot, designed or chosen for its explicit contribution to the desired atmospheric effect.

local 40 man. A bona-fide electrician, not just a crew member who handles lights and LUMINAIRES.

location. Any outdoor setting or interior set that is away from the studio.

location fee. Any money paid for use of a location site and its facilities.

location lighting. Any lighting, exterior or interior, used on locations outside the studio.

location scouting. The searching for sites suitable for a particular production.

location sound. Sound recorded outside the studio.

log. 1. A detailed record of activities during filming on a motion picture. **2.** A detailed record of camera and sound-recording activities during filming. **3.** A daily record of a station's overall broadcast operation required by the Federal Communications Commission.

logo, logotype. A concise graphic symbol used to identify a manufacturer or a product; also used to identify the television networks (the stylized N for NBC, for example) and many local channels.

log sheet. One page of a detailed LOG (def. 1, 2, 3).

long. Referring to a television program that is running beyond its designated time slot.

long focal-length lens, long lens. See TELEPHOTO LENS.

long form. See MINISERIES.

long pitch. The PITCH used on film prints to counteract slippage in the printer.

long shot. 1. A camera shot in which the subject is seen at a distance. **2.** A shot that includes all of a subject and part of the detail of the scene.

loop. 1. A length of film or tape that has been end-spliced to produce continuous projection. **2.** A slack section of film used in any threaded film mechanism (camera or projector) to prevent its being strained or torn by the intermittent CLAW movement.

loop film. See LOOP.

looping. The process of recording dialog to conform with the previously recorded lip movements of the performer. Sometimes the language is not the same as the one used in the original film. (Also called *dubbing.*)

loop printing. Printing made from LOOPS, frequently used for special effects.

loose. The framing of the subject of a shot designed to leave space between the subject and the edges of the frame.

loose gate. A projector GATE that is deliberately loosened in order to minimize film damage.

loose shot. See LOOSE.

LOP (least objectionable program). A theory concerning television audience choices, that home viewers choose not what they would really want but the preferred of available programs, all of which may be mediocre or poor.

lose the light. Referring to outdoor filming that has to be discontinued due to loss of natural light.

lose the loop. To accidentally tighten the LOOP in the slack section of film between the projector picture GATE and the SOUND HEAD so that synchronization is lost.

lot. The area around a studio, and owned by it, in which outdoor sets are constructed for filming or on which filming can occur.

loudspeaker. See SPEAKER.

low angle. A camera position from which the shot is made from beneath the subject with the camera pointed upward toward it.

low boy. A very low TRIPOD mount used for special camera shots.

low-budget production. A motion picture made with limited funds, usually a B PICTURE or a SECOND FEATURE.

low contrast. Muted colors in a film or a long gradation of tones from white to black.

low-contrast filter. A camera lens used to mute colors.

low-contrast original. An original REVERSAL film from which prints having a good projection contrast are made.

low-frequency distortion. Distortion below 15.75 kHz.

low key. Referring to **1.** pictures in which shadowy areas or lower gray scale tones are emphasized; **2.** dim illumination of the subject.

low-key lighting. Any lighting that produces LOW-KEY images.

low-noise lamps. Special lamps used to reduce noise in the audio system while the lamps are in use.

lows, low frequencies. Sound frequencies that begin at 50 Hz.

lumen. One CANDELA of illumination that covers a square foot of surface.

lumens per watt. The number of LUMENS that are produced for each watt in a light source.

luminaire. A lighting device that includes the support, housing, lens, bulb and cable.

M

macbeth. A glass filter that converts a light source to the correct color balance for daylight photography.

MacGuffin, McGuffin. The object in a motion picture on which everyone is concentrated—to steal it, retrieve it, find it, or save it, and so on. For example, the maltese falcon in the picture of the same name or "Rosebud" in *Citizen Kane.* The term was originated by Alfred Hitchcock.

machine leader. A LEADER used in a film processor to pull the film through the processor during its operation.

macrocinematography. The cinematography of objects that are small but not small enough to require a microscope, made with special camera attachments.

macro-focusing telephoto lens, macro-telephoto lens. A specially designed TELEPHOTO LENS used to focus on objects placed close to the camera.

macrolens. A special magnifying lens used for close-up shots.

macrozoom lens. A ZOOM LENS that can bring very close objects into perfect focus.

Madison Avenue, Mad Ave. The street in New York City on which a number of major advertising agencies are located; term used to describe anything in which advertising concepts and practices play a major part.

magazine. A lightproof container for film that feeds the raw film for exposure into a camera and rewinds it after exposure. Magazines are also used for specific printers and optical sound recorders.

magenta. 1. The purplish red elements in a color negative film. **2.** A combination of blue and red lights that results in a purple beam.

magnetic film, mag film. Iron oxide-coated film used to record sound and from which sound can be reproduced.

magnetic film recorder. Any sound recorder that uses perforated magnetic film rather than tape.

magnetic head. An erase HEAD, record head, or playback head that operates in contact with magnetic film or magnetic tape.

magnetic master. A SOUND TRACK that contains a final mixed sound for a motion picture and from which the RELEASE PRINT sound track is made.

magnetic recording. A video and/or sound recording made by passing minute particles of metallic oxide across a modulated magnetic field gap in order to change their polarity.

magnetic sound. Any sound recorded on a magnetic medium.

magnetic stock, mag stock. A quantity of magnetic film received from a manufacturer and stored for future use.

magnetic stripe, mag stripe. An iron oxide stripe that is applied to clear film used for recording a single or mixed sound track.

magnetic stripe sound, mag stripe sound. Any sound that is recorded on a magnetic stripe applied to the sound track of a RELEASE PRINT.

magnetic striping. Applying a magnetic stripe on any kind of film designed to accept it.

magnetic tape. See AUDIO TAPE.

magnetic tape recorder. See TAPE RECORDER.

magnetic track. A magnetically recorded SOUND TRACK that is impressed on a composite film base.

magnetic workprint. A magnetic SOUND TRACK that has been rerecorded and is used in editing and for further rerecording.

magoptical release print. A RELEASE PRINT with both an optical and a magnetic SOUND TRACK.

magoptical sound track. A SOUND TRACK that combines a magnetic stripe track and an optical track.

main title. The title of a motion picture or teleplay as it appears on the screen.

major sponsor. The advertiser who sponsors the most commercials during a program that has a number of advertisers.

makegood. A free rebroadcast of a commercial that has had transmission defects or has been inadvertently omitted.

makeup. 1. A performer's cosmetics. **2.** The application of cosmetics.

makeup artist. A production crew craftsman who applies a performer's makeup.

male. An electric connection plug with prongs which are inserted into a corresponding or FEMALE plug.

maltese cross. A mechanical device that produces the intermittent movement of film through the camera or projector.

manual dimmer. Any light dimmer controlled by hand.

manual override. A camera control used to bypass the automatic exposure control in order to hold the IRIS (see DIAPHRAGM) open at one position.

mark. A chalk mark or square of tape placed on a studio floor to indicate positions for performers or set pieces.

mark it. See STICKS.

marquee. A large sign on the front of a theater, usually hanging above the entrance, on which titles (and occasionally the names of the stars) are shown.

married. Describing the synchronization of picture and sound track on a single filmstrip.

married print. A print that contains both picture and sound track.

master positive. A timed print made from a negative original.

master scene. A scene in which the action is too widespread and complex to be broken into specific camera shots; the camera must be in continual motion. The camera usually remains at a distance (long shot) and medium shots and close-ups are taken later and inserted during editing.

master-scene script. A film script in which the action and dialog are developed in MASTER SCENES rather than in the shot-by-shot technique.

master-scene technique. A director's technique in which the action is covered in a long shot, then repeated for medium and close-up shots, which will be edited into the film. Often several cameras are used to avoid reshooting the scene.

match-action cut, match cut. An action cut between two shots in which the camera position switches from one to the other during the action and the overlapping film is aligned so there is a smooth transition from one shot to the other. See ACTION CUTTING.

match dissolve. A DISSOLVE from one image to another that has a similar content: e.g., a shot of a pine tree in a forest dissolves to a shot of a pine tree with Christmas decorations in a living room.

match-image cut. A CUT from one shot to another shot that is similar in shape: e.g., cut from a beach ball to a spiritualist's crystal ball.

matching, negative cutting. Conforming film NEGATIVE material to the edited WORKPRINT.

matrix. An image filmstrip containing dyed EMULSION that is combined with two other strips on a FILM BASE in order to produce color film.

matte. **1.** A device placed before the camera, used to modify or obstruct light from portions of the action area, or from the film during exposure in the printer. **2.** A dull surface as compared with a slick one.

matte bleed. Some aberration in the matted image that makes the MATTE lines visible.

matte board. A support for MATTES.

matte box. A slotted camera attachment in which MATTES, SUNSHADES, and other optical effects are carried.

matting. The optical or electronic insertion of an image into a background.

matting out. To eliminate an optical element in a shot.

matte line. See BLEND-LINE.

matte ride. Unwanted outlines that appear around a matted element. Cf. MATTE BLEED.

matte scan. A technique used in MATTE photography that uses a computer to direct the camera in precise movements and timing.

matte shot. Any shot in which a portion of the action is screened out so that other action can be inserted later.

MATV (master antenna television system). An antenna arrangement that serves a concentration of television receivers in a specific area.

maximum aperture. The largest opening of a camera lens or a PRINTER HEAD.

MC, master of ceremonies. The host of a broadcast show.

MCU, medium close-up. A camera shot in which the performers are shown from the waist up. Also called *MCS,* medium close shot.

mechanical special effects. 1. Filmed effects, such as fires and explosions, which are created mechanically. **2.** Special effects in which mechanical creatures or objects (such as cars without drivers) are shown as full-scale or larger-than-life models, usually manipulated by wires.

media department. A division of an advertising agency responsible for recommending the purchase of broadcast time to its clients.

medium. The way in which an advertising message is communicated to the audience.

medium lens. A lens with a FOCAL LENGTH that is approximately the normal focal length for the particular format used.

megahertz (MHz). One million HERTZ.

megawatt, MW. One million watts.

melodramatic film, melodrama. A dramatic film in which the emphasis is on sensational plot developments and in which emotions are exaggerated and sentiment (or sentimentality) is magnified.

memomotion. A photographic technique used to represent an extremely slow process at normal projection speeds. See TIME LAPSE CINEMATOGRAPHY.

memorandum agreement. An informal and tentative contract that may be considered valid during the production of a film: e.g., a simple written statement that A has agreed to employ B for X number of dollars to write a specific film script, without details concerning credits, travel expenses, etc.

men's costumer. A production crew member who obtains and cares for the male actors' costumes.

Mercer clip. Manufacturer's name for the plastic clip used to join film ends during assembly.

meshbeat. A watered effect created by undesirable linear qualities of color television picture tubes.

metal halide. A mercury ARC LIGHT that uses AC current only.
meter. A device installed in certain selected television homes in order to record program preferences of the viewers.
method acting. The introspective interpretation of a role by an actor who attempts to transfer his/her personal experiences to and identify closely with the characters being portrayed.
metteur en scène. A motion-picture director who is responsible for all the filmed elements that appear in the completed production. (French: "he who puts [a production] on stage.")
Metro. Abbreviation of the film production company, Metro-Goldwyn-Mayer. (Also called MGM.)
MHz. See MEGAHERTZ.
microfilm. Film used to record printed or written material reduced in size on separate frames so they can be projected for viewing one at a time.
microphone, mike, mic. An electroacoustical device that transduces sound waves into electrical impulses. See TRANSDUCER.
microphone boom. A pole on which a microphone is attached so that it may be extended over the action area.
microphone input. A connection on a sound equipment device into which a microphone can be attached in order to feed the sound into the equipment.
microphone pickup pattern. The three-dimensional area in which a microphone picks up sound effectively.
microphone placement. The position to be occupied by the mike for most efficient sound pickup during shooting.
microphone presence. An undesirable sound that occurs when actor and microphone are too close together and his/her breathing or slight lip sounds are audible.
microphone shadow. The shadow of a microphone caught by the camera and seen on the set.
microphotography. The photographing of small objects requiring optical enlargment to bring out visible details.
microprocessor. A minuscule SEMICONDUCTOR chip that contains more than 5,000 transistors.
microsecond. One-millionth of a second.
microwave. A broadcast system without cables used to transmit both audio and visual signals through highly directional radio beams that extend up to fifty miles, depending upon the topography.
mid-bass. The standard audio frequency range, from 60 to 240 Hz. See HERTZ.
mid-range. The standard audio frequency range, from 240 to 1,000 Hz. See HERTZ.
mid-treble. The standard audio frequency range, from 1,000 to 3,500 Hz. See HERTZ.
mike. See MICROPHONE.

mike man. A production crew member who is responsible for the placement of the mikes.

milk sweep. A small translucent white BACKDROP used to eliminate any visual frame of reference.

miniature. A set or object (such as a boat or train) constructed on a small scale to be photographed so that it appears normal in size.

minicam. A lightweight hand-held camera.

minicam van. An electronically equipped vehicle that contains a portable television "station," complete with control room, and is used to film and record remote broadcasts at the scene of a news story outside the studio.

miniseries. A television series presented in several two-hour weekly segments during a single season, usually the serialization of a novel such as *Roots* and *Shogun.*

minor character. An actor who plays a small part in a film; the role itself.

mired, microreciprocal degree. A numerical value used to indicate the color temperature of a light or a filter. Obtained by dividing 1,000,000 by the appropriate Kelvin value.

mirror ball. A suspended reflecting globe that is composed of many tiny mirrors; light focused on it as it revolves creates thousands of moving reflections.

mirror image. Any image reflected in a mirror, used in film photography for special effects.

mirror shot. 1. A shot of an actor as seen in a mirror. **2.** A means of doubling the depth of a shot by aiming the camera at a large mirror. **3.** A shot made with a mirror in which parts are transparent in order to achieve a ghostly effect.

mirror shutter. A reflex shutter system that enables the cameraman to view the shot as it is being made.

miscast. Referring to an actor used in a role for which he/she is not suitable.

mis-en-scène. The act of combining all of a scene's elements (settings, costumes, lighting, action, etc.) in order to achieve the ultimate desired effect.

missile-tracking camera. A motion-picture camera placed on an electrical mount used to photographically track the path of a missile.

miss the mark. An actor's move to a set point in which he misses the mark on which he is supposed to stand.

Mitsubishi. A Japanese electronics firm.

mix. 1. To combine various sources of sound in a single recording. **2.** The recording itself. **3.** A musical session in a rerecording studio. **4.** To gradually merge the end of a shot into the beginning of the next one.

mix cue sheet. See CUE SHEET.

mixer. 1. The technician who is responsible for rerecording and mixing sound. **2.** Sometimes used to refer to the MIXER BOARD.

mixer board. A mix-control console used in the mixing of various audio elements.

mixing. The process of combining sound tracks in order to achieve an acceptable single sound track.

mixing studio. The facility where the audio elements are mixed into a final sound track.

MLS (medium long shot). A camera shot in which the image size appears to be between that of a medium shot and a long shot.

MNA (Multi-Network Area). The NIELSEN group of thirty major market areas in which programs of the three major television networks are transmitted to the local stations.

mobile unit. The equipment, including vehicles, used to record and transmit electronic signals from a location outside a studio.

mob scene. A large group of extras acting as a crowd, frequently an angry or excited one.

mock-up. A full-scale model of a set or object.

mode. The electronic setting that activates specific circuits.

model. A miniature replica used in SPECIAL-EFFECTS cinematography, such as the representation of the solar system in *Star Wars.* Miniatures can be combined with shots involving live actors by OPTICAL PRINTING techniques.

modeling. The illumination of a subject that emphasizes its contours.

modeling lights. LUMINAIRES used to create both shadows and highlights on objects and actors in the action area.

model sheet. Animated drawings in which the characters are shown in various poses to be used as models for animators.

model shot. A camera shot in which a model or models are used.

modulation. 1. The process of altering the level of sound. **2.** Recorded audio signal patterns.

module. An electronic part or component that is interchangeable with other parts or components.

moiré. 1. An undesirable "watered" optical effect caused by a set of closely spaced lines inaccurately imposed over another set. **2.** Disturbance in the picture created by interference beats of similar frequencies.

Molevator. A power-operated extension stand, usually from six to fourteen feet high; for large spotlights.

monitor. 1. To oversee and control various elements of a motion picture during filming. **2.** A television receiver that is wire-connected to a transmission source and used for in-station viewing.

monitor speaker. A loudspeaker used for listening to all audio elements during recording and mixing sound.

monitor viewfinder. A VIEWFINDER in which the image is seen, without an EYEPIECE LENS, on a small screen or a segment of ground glass.

monochromatic. Referring to images seen in tones or gradations of a single color or hue.

monochrome. An image seen in black, gray, or white.

monochrome transmission. A signal wave that represents the components of brightness but not the color values of a picture.

monopod. A single-leg support for a camera.

montage. The visual juxtaposition of a series of short shots, often SUPERIMPOSED, to create an impressionistic effect, usually to bridge a time lapse by indicating the events that occurred within it.

mood music. Background music used to elicit a particular emotional response from the audience to what is being shown on the screen.

MOS. An indication that no sound is to be recorded during the filming of a scene. Attributed to a German-speaking director during the early days of sound motion pictures, said to have ordered "Mit-out sprache" (without sound).

mosaic. The storage surface of a camera pickup tube that is scanned by an electronic beam. See PICKUP (def. 4).

motif. A theme, either for a film plot or a musical composition.

motion picture. A succession of still pictures or images that appear to be in motion when projected on a screen or television tube face.

Motion Picture Association of America (MPAA). An organization of the principal producers and distributors in the United States, founded for the purpose of establishing moral and artistic standards in the industry; affiliated with ASSOCIATION OF MOTION PICTURE AND TELEVISION PRODUCERS.

motion-picture camera. A box equipped with lens, film advance system, shutter and viewfinder in which motion-picture film is exposed.

motion-picture film. Film that is sensitive to light, used in a motion-picture camera.

Motion Picture Production Code. A list provided to motion-picture producers that indicates subject matter and vocabulary that may not be used in films, adopted by the Motion Picture Producers and Distributors of America in 1930, superceded by the rating system in 1966. (Formerly known as the Breen Code and the Hays Code).

motor cue. The first of two small circles shown on the screen near the end of a reel to alert the projectionist that it is time to start the motor of the projector holding the next reel.

mount. A stand on which a camera is placed for elevation.

movie. A motion picture, a film.

moving backgrounds. Filmed action that is projected against a rear screen.

moving camera. Referring to a shot made while the camera position is being changed.

moving shot. A shot made from a mobile camera that accompanies the action as it moves from place to place.

Moviola. 1. An editing console that produces a small image of the picture track with its accompanying sound, used in the editing process. **2.** Any similar viewing instrument.

MPA. A television commercial that advertises *m*ultiple *p*roducts.
MS (medium shot). A camera shot in which one or more performers are shown from the waist up.
motivation. The plausible reason(s) for the behavior of an actor in a role, as shown in the development of the plot through action and dialog.
mug. An actor making unnecessary facial contortions, to attract attention.
multicamera. Referring to filming and taping done simultaneously from two or more camera positions.
multilayer color film. Film that has two or more layers, each of which is sensitive to a different color or hue.
multiplane. Describing animation artwork that is layered so that animated figures in the foreground move in front of scenes and figures in the midground and background.
multiple exposure. The addition of images to a filmstrip after exposure, made by two or more exposures on a single series of film FRAMES.
multiple-frame printing. Printing a single FRAME several times.
multiple-head printer. A PRINTER that contains three HEADS, one from the *A roll,* one from the *B roll,* and one from the SOUND TRACK. See A AND B ROLLS.
multiple image. A frame in which several images are combined, all from different sources and none superimposed.
multiple interference. Cancelled sound frequencies that result from two microphones placed too close together.
multiple printing. The printing of several images from different filmstrips onto a single strip.
multiple-screen presentation. The presentation of two or more films on two or more screens in the same theater, or on sections of a single large screen.
multiplex. A single conductor that accepts or transmits more than two signals simultaneously.
M and E (*m*usic and sound *e*ffects). The film sound tracks, exclusive of dialog, which are needed for the dubbing of foreign languages and other voice dubbings.
musical comedy film, musical, musical comedy. A full-length film in which emphasis is on the music and dancing rather than the dramatic content.
music track. A sound track on which music is recorded for a film.
mute negative. A picture NEGATIVE without the sound track.
mute print. A POSITIVE film print without a sound track.
Mutoscope. An early peep-show viewer in which the illusion of motion (however jerky) was obtained by rapidly flipping cards that contain photographed images.

N

NAB (National Association of Broadcasters). An organization that represents radio and television broadcasters, providing guidelines for acceptable programming and advertising practices.

NAB Code. Standards governing both programming and advertising among the members of the National Association of Radio and Television Broadcasters.

NAB curve. The standard for audio playback equalization.

NABET (National Association of Broadcast Employees and Technicians). The trade union for broadcast technicians.

NAEB (National Association of Educational Broadcasters). A membership organization of television station operators.

Nagra. A small portable audiotape recorder used on locations outside the studio.

narration. 1. Any commentary that is heard during a film from an off-screen voice. **2.** Commentary delivered on camera.

narration script. The script prepared for use by a narrator.

narrative. In a film script, the description of the scene or shot, including action, that precedes the dialog that accompanies the shot.

narrator. 1. An off-screen commentator, usually one who provides needed exposition. **2.** A performer who delivers commentary during a television broadcast.

narrow-gauge film. Any film that is less than 35mm wide.

NASA (National Aeronautics and Space Administration). The United States government agency that administers the communications satellite program.

National Academy of Television Arts and Sciences. Organization of professional television actors, directors, photographers, producers, writers, etc.; sponsor of the annual ceremony in which EMMY awards are given for excellence.

National Education Television. See NET.

National Public Affairs Center for Television. See NPACT.

naturalism. The dramatic and objective use of realism in which characters appear to be at the mercy of natural (usually brutal) forces such as environment, heredity, or evolutionary factors.

natural light. Any light, such as sunlight, that is not artificial.

nature film. A motion picture about plants and animals in their natural environments.

NBC (National Broadcasting Company). One of the three major broadcasting television networks, owned by the RCA Corporation.

NCTA (National Cable Television Association). The membership organization of cable system operators.

needle. 1. A meter dial indicator. **2.** A stylus that is used in a record-player arm to track the record grooves.

needle drop. The one-time use of a composition of licensed *stock music.* See LIBRARY MUSIC.

negative. 1. A black and white film image with tonal values that are the opposites of those in the original material. Occasionally negatives are projected on the screen to create special effects. **2.** A color film image with color values that are complementary to those in the original material. **3.** A film having negative images. **4.** A television signal that has tonal and/or color values opposite or complementary to those of the original material.

negative cost. The expenses incurred in filming a picture prior to its release exclusive of distribution and exhibition costs.

negative cutting. Conforming the film negative material with the edited WORKPRINT.

negative film, negative stock. Especially designed film used to produce a fine negative image when exposed or processed.

negative image. 1. In black and white, an image on which the light and dark areas of the original subject are reversed. **2.** In color, an image in which the color values of the original images emerge as complementary values, in addition to the reversal of the black and white tonal values.

negative numbers. The manufacturer's numbers that appear on the edge of film.

negative pitch. The lesser distance between camera film stock PERFORATIONS as compared with those on print film stock.

negative pulling. See NEGATIVE CUTTING.

negative scratch. Any scratch that appears on film or, as a result of it, on the print made from the film.

negative stock. See NEGATIVE FILM.

Nemo. The origination point for any remote broadcast signal. (An acronym from an early telephone company designation, *N*ot *E*manating *M*ain *O*ffice.)

neorealism. 1. A film style developed in Italy after World War II in which man's struggle against an indifferent society is the predominant theme; often shot with amateurs (real persons in real situations) at the location where they lived and worked. **2.** In American films neorealism focused on the Depression years (*Grapes of Wrath*) and on the World War II era (*Story of G.I. Joe.*)

net. A metal or gauze DIFFUSER used on a spotlight to soften the illumination on a set.

NET (National Education Television). The programming organization for educational television.

network feed. Referring to programs that originate in New York, Chicago, and Los Angeles that are fed to stations throughout the United States via AT&T cables and MICROWAVE links.

neutral density filter (ND filter, ND). A gray lens filter used to reduce the transmitted light so that colors, contrast, and definition are not affected.

newscaster. A person who presents the news on a broadcast program.

newsfilm. Film of spot news, sports, or feature stories shown during regular news broadcasts.

New Wave (*Nouvelle Vague*). Feature films, beginning in the late 1950s, produced by French directors with a strong improvisational style over which the director had supreme creative control and on which his personal stamp was recognizable. Plots were loosely woven, usually around alienated persons in an unstable world. Usually filmed on actual location, rather than on studio sets or lots.

NG takes, NG. *N*o *g*ood shots, those that will be discarded.

nickel-cadmium, ni-cad. A portable rechargeable storage battery.

nickelodeon. A small motion-picture theater in the silent-film era; a nickel was the admission price, *odeon* is from the Greek work for theater.

Nielsen, Nielsen rating. An A.C. Nielsen Company audience survey service using AUDIMETER devices that record the viewing habits of 1,250 sample United States homes. Each rating point is said to represent 833,000 homes, or 1 percent of the 83.3 million United States households that have television (1982). The television networks use the weekly ratings to determine the prices they charge for commercial time. See NTI.

night filter. Filter used to alter the color of a daytime shot in order to present the illusion that it was shot at night.

night-for-night. Night sequences that are shot at night. Cf. DAY-FOR-NIGHT.

nighttime. The television broadcast time period from 7:00 P.M. to 11:00 P.M. or midnight.

nitrogen-burst. The systematic emission of nitrogen into a film-processing bath that produces AGITATION.

nodal point mount. A CAMERA MOUNT designed to permit camera rotation around one of the nodal points of the lens during a panning operation. See PAN.

no fax. A rehearsal for performers during which no technical facilities are used.

noise. Extraneous sound in an audio pickup, usually caused by random energy generated by voltages within some electronic equipment that causes interference in the sound signal.

noncommercial. Referring to any program or broadcast material that does not have commercial advertising.

non-dim circuit. An electrical system that does not provide for dimming, having only an on-and-off function.

nondirectional microphone. A microphone that has an even pickup response in all directions.

nonfiction film. A film that provides information for educational or documentary purposes.

non-hero. The protagonist in a film who does not represent the qualities expected in a traditional hero; often in the end a loser.

nonsynchronous sound, nonsync sound. Any recorded sound made when no camera is operating in sync with the recorder.

nontheatrical film. An educational film made to be shown in places such as schools and churches, other than motion-picture theaters.

normal lens. A cinematographic lens with a focal length approximately twice the diagonal width of the action area.

north. The upper section of an animation field chart.

northlight. A LUMINAIRE used for diffused fill lighting. See FILL LIGHT.

no-seam. A large roll of wide colored paper used for backgrounds in which no visuals are needed.

notch. 1. See CUE MARK. **2.** An EMULSION mark on the edge of filmstrip, used for identification purposes in the dark room.

Nouvelle Vague. See NEW WAVE.

NPACT (National Public Affairs Center for Television). A television program production unit of the federal-government-funded Public Broadcasting System with headquarters in Washington, D.C.

NTI (Nielsen Television Index rating). A rating report that indicates the biweekly size of a television network audience; based on the AUDIMETER records for the 1,250 sample households.

nudie. A film in which the emphasis is on nudity, nearly always female nudity. Usually X-RATED.

null. An area into which the microphone does not carry sounds.

number board. A board briefly held before the camera and photographed before each shot, on which is recorded the film title, the number of the shot, and the number of the TAKE, for editing identification purposes. Usually has CLAPPER attached.

nuts-and-bolts film. A film that places emphasis on practical information and in which style is played down in order to present unadorned facts.

O

O and O's. Broadcast television (and radio) stations that are both owned and operated by one of the three major networks.

oater. A WESTERN motion picture.

objective. The lens in the camera; any optical system with image-forming ability.

objective camera, objective camera angle. A camera angle in which the shot is seen by members of the audience as if they were actually observing the action from their theater seats, such as a straight-on shot of a scene as it might appear on a stage. Excludes long shots, establishing shots, aerial shots, etc.

obligatory scene. A scene that is necessary in order to resolve the plotted problems and conflicts that have preceded it; particularly important in climactic scenes and those of the DENOUEMENT.

observation port. An opening in the camera PROJECTION BOOTH through which the projectionist can see both the screen and the audience.

octopus box. See SPIDER BOX.

off-camera. 1. Referring to anything that occurs outside the area of the action shot. **2.** In television, sometimes used to describe an off-camera (VO) voice.

off-microphone, off-mike. Describing **1.** sound that is beyond the pickup pattern of the microphone; **2.** speech that is deliberately directed away from the mike to simulate a sound heard at a distance.

off-screen, OS. Describing action that occurs out of camera range but is supposed to be occurring nearby.

ohm. A basic unit of electrical resistance.

OIRT (International Radio and Television Organization). A worldwide group that sets telecommunications standards.

OK takes. Takes in which all audio and visual elements are acceptable for WORKPRINTS.

omnidirectional microphone. A microphone capable of picking up sound evenly in all directions.

omnimax. A theater in which are shown films produced with the IMAX technique; has an enormous screen that surrounds the audience and an enveloping sound system.

on camera. Referring to any performer who is both seen and heard on screen.

one-light print. A positive film print made with a single PRINTER LIGHT for all exposures being printed; used for editorial work.

ones. The exposure of one ANIMATION frame for each drawing.

one-shot. 1. A shot in which only one performer is seen. **2.** A single performance not intended for rebroadcast.

1,000-hertz tone. A standard audio reference tone signal. See HERTZ.

one-way set. A film set that contains only one flat background.

on location. Filming that takes place outside the studio.

on mike. Referring to a performer who is speaking directly into the microphone.

on speculation, on spec. Referring to **1.** producing a film for which no sponsor has been obtained, taking the risk of making a profit through rentals and sales; **2.** writing a film script without a contract, taking a chance that it will find a buyer.

on stage. Within camera range.

on the air. Referring to any material that is being broadcast.

on the nose. Describing **1.** filming done at exactly the exposure indicated by the light meter; **2.** perfect timing.

OOP. Out-of-pocket expenses.

opacity. 1. The amount of light absorbed by film emulsion. **2.** Loosely, OPAQUE.

opaque. Describing a surface that does not transmit light.

opaque leader. A section of filmstrip used as a LEADER in which no image appears.

opaquer. An ANIMATION artist who applies paint to the inked-in outlines of drawings/CELS.

open call. A general audition.

open end. 1. A broadcast commercial in which time is allowed for the addition of local material. **2.** A program that has no completion time scheduled.

open mike. A microphone that is in operation.

open reel. A transport system for tape in which there are separate supply and take-up reels.

open up. To enlarge the lens APERTURE of the camera.

operating cameraman. 1. The crew member who runs the camera and controls its angles as it covers the action. **2.** Sometimes used to refer to the theater projectionist.

OPT (Operation Prime Time). A group of independent television stations joined in an effort to provide programming alternative to that offered by the networks.

operations department. The studio group in charge of television program scheduling.

operations sheet. The daily schedule for all broadcast programs.

operator. See OPERATING CAMERAMAN.

optical effects, opticals. Special visual effects that are achieved artificially by means of an OPTICAL PRINTER: e.g., DISSOLVES, FADES, WIPES, FLOP-OVERS, HOLD-frames, SUPERIMPOSITIONS; often used as transitional devices. More sophisticated alterations of time and space are seen in the SPECIAL EFFECTS used in films such as *Star Wars, Superman,* and *Raiders of the Lost Ark.*

optical flop. An image that is reversed (flopped over) in an OPTICAL PRINTER.

optical glass. Special high-quality glass used in camera lenses.

optical house. A business facility in which final film negatives are processed; optical effects and titling are included in the processing.

optical negative. The negative used in the final picture printing.

optical print. Any print made by an OPTICAL PRINTER.

optical printer. 1. A printing apparatus that includes both a projector and a camera, used to produce the final OPTICAL NEGATIVE. **2.** Photographs images from one film onto another for SPECIAL EFFECTS and trick work.

optical printing. Printing done on an OPTICAL PRINTER rather than on a CONTACT PRINTER.

optical reduction. Printing from one film to another smaller film size: e.g., from 35mm to 16mm.

opticals. See OPTICAL EFFECTS.

optical sound. Sound that is recorded or reproduced from a photographic sound track, rather than from magnetic film, tapes, or records.

optical sound recorder. A recorder that produces a photographic sound track.

optical sound track, optical track. 1. The final sound track PRINTING NEGATIVE. **2.** A film's patterned photographic strip that is converted into sound by an EXCITER lamp beam and a photoelectric cell.

optical system. The components of a motion-picture camera that produce visual images on film.

optical track. A final sound track PRINTING NEGATIVE.

optical transfer. The transfer of sound from a magnetic tape track to an OPTICAL SOUND RECORDER, to make a negative or positive film sound track.

optical viewfinder. A device containing an OBJECTIVE lens and an EYEPIECE lens through which the camera operator can accurately frame an action area.

optics. The components of an OPTICAL SYSTEM.

option. A formal agreement in which the prospective buyer, for a fee, has exclusive rights to a property for a stated period of time, after which either he buys it outright or the rights return to the original owner.

original. The initial photographic negative or videotape recording, prior to duplication and/or editing.

original screenplay. A script that is the product of a screenwriter's imagination rather than an adaptation of a novel, short story, or play.

origination. The geographic point at which a network program is fed to its member stations, usually New York, Chicago, or Los Angeles.

orthochromatic film, ortho film. Film treated with an emulsion in which sensitivity is limited to blue and green areas of the spectrum, excluding orange and red, used to achieve a correct color tone.

Oscar. The nickname for the statuettes awarded annually by the ACADEMY OF MOTION PICTURE ARTS AND SCIENCES for major achievements among persons in all branches of the film-making industry. (Said to have been named by Bette Davis for her husband, Harmon Oscar Nelson, when she won her first statuette for her role in *Dangerous* in 1935.)

oscillator. An electronic device that produces specific frequencies.

oscilloscope. See SCOPE.

OTO (one time only). A broadcast commercial scheduled to be aired only once.

out-cue. The last words of a CUE line.

outgrade. The elimination of a performer from a script when it is edited or revised.

outline. A plot summary for a motion picture or television film, to be presented to prospective buyers.

out of character. Referring to **1.** a performer who does not sustain the characterization of the role he/she is playing; **2.** dialog or action not suitable to the role.

out of frame. Referring to **1.** anything outside the action area or camera range; **2.** a faulty setting of the framing mechanism, causing portions of two frames to be visible on the screen.

out-of-sync. Referring to audio and visual elements that are not aligned: e.g., the voice sounds do not coincide with the lip movements.

out-take. A shot that is deleted from a film during the editing process.

overcrank. To operate a motion-picture camera at a frame speed faster than normal, thus producing a slow-motion effect when it is projected at normal speed.

overdevelop. To develop a filmstrip longer than is necessary, which increases the fogging effect; sometimes done intentionally.

overexpose. To expose film longer than necessary by using a too-slow shutter speed or an overwide lens aperture; results in a dark negative and light print.

overhead, overhead expense. Ongoing film company expenses, such as utilities and building maintenance, which are not connected to any specific production.

overhead shot. A camera shot made from a position above the action area.

overlap. 1. The repetition of some action at the end of one shot and the beginning of the next shot so that a MATCH-ACTION CUT can be achieved. **2.** To sustain a sound from one shot into the next.

overlap dialog. A line of dialog that interrupts and is spoken during dialog already in progress, often used to achieve an effect of realism in the dialog.

overlapping and matching action. 1. Repeating part of the action that ends one shot at the beginning of the next shot. **2.** Photographing the action in two adjoining shots with two or more cameras so the overlaps can be matched to produce a smooth cut on action.

overload. To use too much power or signal in equipment that is not capable of handling it without distortion.

overpower. To overcome the harsh effect of fluorescent lights by the use of sufficient light at the proper color temperature.

override. See MANUAL OVERRIDE.

overscale. Performers' fees that are above union minimums.

overscan. A monitor control term used to indicate that the picture seen on station monitors exceeds in area the picture seen on home sets, the larger picture used to monitor vertical and horizontal requirements.

overshooting. Filming more footage than is needed, usually to provide for choices during editing.

over-the-shoulder shot (OSS). A camera shot made from behind an actor, sometimes including a portion of his/her shoulder and head, with the camera directed on the spot at which the actor is looking.

oxidation. The passage of oxygen into a chemical compound, causing a reduction of strength in various photochemicals.

oxide. The microscopic oxidized metallic particles that are base-coated to form magnetic tape or film.

P

P. A mark made on a TAKE to denote that it is to be printed by the film laboratory.

PA (public-address system). A combined microphone and loudspeaker system used at public meetings to amplify sound.

package. **1.** A combination of two or more persons or properties (such as scriptwriter and star, or director and star, and the rights to a best-selling novel), which is presented to a potential backer to make a film more marketable. **2.** A prepared television program or series offered for sale.

package plan. A sale-priced combination offer for SPOT commercials to be presented on a weekly or monthly basis by a television station.

packager. The company that produces a television program or series PACKAGE. For motion pictures, the packager is usually an AGENT.

pad. **1.** To insert extra material to fill in the allotted time. **2.** The material itself. **3.** In film scripts, to enlarge an actor's part by increasing his/her lines.

paddle plug. A flat electric plug used on a stage.

painted matte. A MATTE on which images are painted in order to fill out an action area in which the scene is incomplete.

painting on film. Painting colored images on clear or translucent film frame-by-frame or over sections of the film.

paint pots. The color control RHEOSTATS on a television studio console.

pan, pan shot. A camera shot in which the camera swivels in a horizontal plane around its vertical axis from a fixed position to scan the scene before it. The camera does not focus on any single action but occasionally is used to follow the action or to give a *pan*oramic view of the scene.

pan-and-tilt head, pan-tilt head. A camera mount that assures the smooth movement of a camera in a PAN SHOT; can be attached to DOLLY, TRIPOD, or MOUNT.

pancake. 1. Water-soluble makeup and cosmetics base used by performers. **2.** A support box, lower than an APPLE BOX.

panchromatic film, pan film. Black and white film that is sensitive to all colors of the visible spectrum.

panchromatic master, pan master. A black and white positive made from a color negative, in order to make a black and white duplicate negative.

panhandle. A lever attached to a pan-tilt camera HEAD to control its movement.

panorama. A wide-angle shot in which the entire scene is shown or is revealed by a PAN SHOT.

pan shot. See PAN.

panstick. Performer's makeup, grease-based, unlike the water-soluble PANCAKE makeup.

pantograph. An overhead adjustable hanging support for a LUMINAIRE.

paper print. A motion-picture print made on paper coated with an EMULSION.

parabolic, parabolic spotlight. A spotlight that projects a narrow beam of light.

parallax. The angle of divergence between the viewfinder objective and the camera lens, which might cause framing errors.

parallax correction. Any method used to correct errors caused by PARALLAX.

parallax error. 1. An uncorrected VIEWFINDER parallax. **2.** The framing error caused by such omission.

parallel. A platform used in a studio or on location to elevate the camera and crew above the action area for high-angle shots.

parallel action. A series of shots of two or more events shown alternately, to convey to the audience that they are taking place simultaneously: e.g., shots of a child fighting for its life in a hospital are alternated with shots of a pilot flying through a storm to deliver the life-saving serum.

parallel editing. INTERCUTTING two or more shots to show events taking place simultaneously.

parallel development. Two or more plot developments shown through CROSS-CUTTING.

parody. A humorous or satirical film or television skit based on a serious literary work whose style it imitates.

part. An actor's role.

participation. Referring to a broadcast program that accepts SPOT commercial insertions that are noncompetitive.

party line. See PL.
passive. Describing any equipment that cannot generate or amplify power.
patch. Electronic circuit connection made by plugging short cables into panel connections.
patch bay, patchboard, patch panel. A mounted set of circuit connectors.
patch cable, patch cord. The short cable used to connect one electrical component to another on a PATCH PANEL: has MALE connectors at each end.
patching. To connect PATCH CABLES.
patch plug. A FEMALE electric cable connection mounted on a console.
path. The route traveled by an electronic signal.
pathos. Any element in a film that has the power to evoke feelings of compassion or pity in the audience.
payola. Illegal or unethical payments made to game contestants by a television station or to broadcast personnel by persons desiring favored treatment for a product.
pay-per-view. Referring to motion pictures released by studios to cable or subscription television services for subscribers who have requested them and pay an extra fee to receive them.
pay television. Cable services that charge monthly fees to home subscribers who have special transmitting equipment installed by the service.
PBS (Public Broadcasting System). An interconnected television system, partially funded by the United States government, which distributes programs to more than 225 noncommercial stations. (Sometimes identified as "*P*lenty of *B*ritish *S*hows" because of a preponderance of programs from the British Broadcasting Company [BBC] and other British sources: e.g., the *Upstairs, Downstairs* and *Brideshead Revisited* series.)
PC spot. See PLANO-CONVEX SPOTLIGHT.
P.D. (public domain). Any creative work that is not copyrighted or on which the copyright has expired and, therefore, can be used without payment to the author or composer or to his/her estate.
pea bulb. A tiny lamp inside a motion-picture camera which produces a FLASH FRAME edit cue.
pedestal. A DOLLY support for a camera or LUMINAIRE.
peg bar, peg board. See ANIMATION BOARD.
pellicle. A sheet of film that has a slick or mirrored surface.
pencil. A rough animation sketch made on white paper and photographed in order to check the speed and movement of cartoon characters before the CELS are painted.
pentode. An amplifying vacuum tube that contains three variably charged wire-mesh grids that control the electron flow between the CATHODE and the positive PLATE.
perambulator. A mobile platform support for the BOOM operator and the microphone boom.
perceptual film. See STRUCTURAL FILM.

perforations. Sprocket holes in motion-picture film or magnetic film (and some magnetic tape).

permanent set. See STANDING SET.

persistence of vision. The phenomenon of image retention caused by the time-lag effect between visual stimulation and the loss of response to that stimulation. All film illusion is based on this persistence that occurs when static images, each slightly changed from the preceding one, are displayed faster than the brain or the optic nerve can comprehend or react to them (more than ten to fifteen times per second).

perspective. The illusion of depth as objects recede into the distance from the point of view of the audience; the angle at which the audience sees things.

perspective distortion. The apparent distortion of an image when it is viewed from any point other than the center of perspective: e.g., when seen through a wide-angle lens an object seems more rounded than it actually is; when seen through a telephoto lens an object seems closer to objects in the foreground than it actually is; any action moving directly toward or away from the camera seems much slower than it actually is.

PG-rated. The label given by the Code and Rating Administration rating board of the Motion Picture Association of America to motion pictures for which parental guidance is advised for young audiences.

phase. The coincidence of COLOR BURST and reference signal.

phase distortion. Changes in the desired picture color.

phasing. 1. The standard alignment process for television cameras and videotape recorders. **2.** A loss in the quality of transmitted sound when two microphones are placed too close together.

Phonoscope. A 1927 invention that recorded television signals on a wax disk.

phosphorescence. The production of heatless light through the absorption of energy.

phosphors. The chemical coating inside a picture tube which lights up when exposed to electron beams.

photoconductor. A conductor that permits a variable current flow when exposed to light.

photoelectric cell, photocell. A device that converts light variations into corresponding variations of electric impulses that can be used as an audio signal.

photoelectric process. The process of changing light into energy (electric voltage), which is the basic function of LIGHT METERS.

photoflood. An incandescent light bulb with a high wattage.

photogenic. A term used to describe a performer whose features photograph unusually well.

photography. A process by which an image is formed on a sensitized surface through exposure to light or other radiant energy.

photo matte. A MATTE that is part of a photograph.
photometer. A photographer's LIGHT METER.
photometry. The science of measuring light.
photoplastic. Referring to a technique for recording images with light and heat on special plastic film.
photoplay. **1.** A script for a motion picture. **2.** The motion picture itself.
photoresistant. Describing material that reacts to light by hardening.
photosensitive. Describing material that reacts to the presence of light.
Photostat. A photographic copy; may be enlarged or reduced in size from the original.
picaresque. Referring to a film, usually historical, in which the hero is a likeable scoundrel: e.g., *Tom Jones*.
pick up. To insert a shot in a film.
pickup. **1.** A receiver mechanism that includes the needle, needle cartridge, and arm used to play back sound from a record/disk. **2.** A remote television broadcast. **3.** The sensitivity area of a microphone. **4.** The television camera tube that converts optical images into electric impulses through an electronic SCANNING process.
picture. **1.** A filmstrip containing images on sequential frames. **2.** A motion picture. **3.** The image area of a film. **4.** The portion of the composite television video signal above the blanking signal that contains the image information.
picture head. The part of the projector that projects the film images on the screen.
picture safety. Referring to the picture-tube area that contains all image detail within the mask edges on receivers that might be overscanned.
picture tube. The CATHODE RAY component (tube) of a television receiver which converts electronic signals to fluorescent optical images by variations in the scanning beam intensity.
piggyback. A broadcast commercial in which different products of the same company are advertised.
pigtail. A three-wire electrical hookup used to provide power from 230-volt lines.
pilot. The initial program of a dramatic/comedy television series, produced to be presented to potential sponsors for series funding.
pilot pin registration. The positioning with PILOT PINS of camera frames in relation to the sprocket holes so that the film advances.
pilot pins. The tiny prongs that engage the film sprocket holes in order to hold the frame steady in the camera or projector GATE.
pilot print. The initial film PRINT.
pincushion distortion. A kind of image distortion caused by a lens aberration in which the corners of square objects appear to be extended.
pinhole. A clear dot seen in the developed EMULSION of a film, frequently caused by insufficient agitation of the processing fluids.

pin rack. A row of hooks on which perforated filmstrips are hung.

pipe. A wire hookup for television transmission.

pipe boom. A microphone BOOM made with two fitted sections of tubing that can be elongated.

pipe grid. See GRID.

pirate. 1. To illegally copy or broadcast a transmission signal. **2.** To tap cable television transmissions without subscribing to the service. **3.** To make an illegal copy of a film or print.

pirated print. Any film print made illegally from another print or original. **2.** A duplicate print of a film.

pistol grip. A handle that provides an easily controlled grip, attached to the bottoms of hand-held cameras or microphones.

pitch. 1. The standard distance between the leading edges of the sprocket holes on film to be used as print stock. **2.** Sound wave frequency.

pix. Abbreviation for pictures.

pixels. The electronic television *pi*cture *el*ement*s* that compose the SCANNING LINE; transmitted at the rate of 8½ million per second.

pixillation. A film animation technique that uses rapid cutting between still shots to give an appearance of movement.

pixlock. The corrected color synchronization between two videotape recorders.

PL (party line). A wired communication system used on the television studio set.

plain lighting. Artificial light that is used to simulate the normal angles of sunlight.

plan. A scale drawing of the set as viewed from above.

plano-convex spotlight (PC spot). A spotlight that uses a lens that is flat on one side, and convex on the other; produces a narrow beam of light.

plant. A plot device in which an object or an idea is inserted in a scene casually in order to make dramatic use of it later on in the film.

plate. 1. A sheet-glass support that contains a photographic image. **2.** A rewind disk that supports film being wound on the core. **3.** An element of a positively-charged vacuum tube. **4.** The dimensions of a console editing machine according to the number of plates it can accommodate, e.g., two-plate, four-plate, six-plate. See FLAT-BED EDITING MACHINE.

platen. A clear glass plate that is placed on animation CELS to hold them flat during photography.

playback. 1. The reproduction of recorded material, often used on a sound stage when performances of singers and dancers are filmed under conditions of imperfect acoustics. The action is later synchronized with the original recording to produce an acoustically perfect sound. **2.** Any device used for such reproduction.

playback track. A SOUND TRACK used as a running CUE by performers during action shots while the sound is being played.

players. Principal performers in a broadcast commercial.

Players Guide. A directory of performers.

playlist. A list of musical recordings to be played during a broadcast program.

plot. 1. The dramatized development of the basic story idea used in a motion picture or television film. **2.** To create such a story, sequence by sequence.

plot gimmick. See GIMMICK.

plot line. 1. The story line of a film. **2.** A line of dialog that essentially advances the plot.

plot plant. See GIMMICK.

plug. To promote a product or performer on the air.

plugging box. A device used as an interconnector for stage lights.

plugola. The broadcast promotion of a product in exchange for merchandise; the term usually refers to an excessive degree of the practice.

Plumbicon. 1. Trade name for a color television camera pickup tube that has a lead-oxide target surface coating. **2.** A camera that contains such a tube.

pocket. A permanent FEMALE electrical outlet used on studio sets.

point-of-view shot (POV). A shot made from a camera position that approximates that of the performer so that the audience sees what the actor is seeing.

polarity. The positive or negative characteristics of a black and white television image.

polarized lens. A lens that has a POLARIZING FILTER attached.

polarizing filter. A lens filter in which optical slits permit polarized light to pass through; or can also be used to reduce the amount of polarized light, depending upon the angle of the slits.

polarized light. Light that passes through lenses or plates of millions of crystals and blocks all waves except those vibrating in the same plane; minimizes glare and reflections.

Polaroid filter. The trade name for a lens filter that polarizes light in order to eliminate glare or unwanted reflections from objects during shooting.

Polaroid shot. A still picture made with a Polaroid camera at the end of a take, used by director and cameraman to check the position of the performers for reference when shooting continues.

Polecat. The trade name for a telescoping support for LUMINAIRES and other equipment.

polyester. Film and tape base made of polyethylene glycol terephthalate.

pool-hall lighting. Illumination from a single source, usually one that is visible and hanging over the middle of the set: e.g., a light bulb dangling from a cord.

poop sheet. A fact sheet used by a television announcer as a basis for AD LIB speaking.

pop. **1.** An explosive sound made by a voice during recording, usually on the letter *p*. **2.** Contemporary music.

pop filter. An internal device in a microphone that limits or eliminates the POP sound.

pop-off, pop-on. The instantaneous addition (or substraction) of optical images on a frame at precisely the point intended.

pornography film, porno film (flick, movie). A film in which explicit sex scenes dominate the action.

portapak. A small portable camera or recording deck ensemble operated by batteries.

position. **1.** The spot on a television program occupied by the commercial. **2.** The location of recorded material on sound tape.

positive. **1.** A film with positive images prepared for projection. **2.** A film in which the color and tonal values are the equivalent of those of the original subject.

positive distortion. See PINCUSHION DISTORTION.

positive image. A film image with tonal and color values equal to those of the original subject.

positive pitch. The standard difference in PERFORATION spaces and sizes on print film.

positive print. A POSITIVE-IMAGE film that has been processed.

positive scratch. The black mark on a print made by a scratch on the original positive.

positive sound track. A SOUND TRACK with a clear track pattern.

post-dubbing. Any dubbing that is done after the film is completed.

post-production. Any work done on a film after it is completed.

post-recording. Any recording done, usually to match voices and images, after photography on the film has been completed.

post-scoring. Recording music to be used in the edited film.

post-synchronization. The synchronization after the film has been shot between the performers' lip movements and the sound, either in the language of the original film or in another language that is recorded and added to the film.

pot, potentiometer. A round console RHEOSTAT used to control audio or video levels.

POV. See POINT-OF-VIEW shot.

power. The wattage output of a broadcast transmitter.

power cable. Any cable used to connect electrical equipment to a power source in the studio or on location.

power pack. A rechargeable portable battery power source for camera or recorder.

power zoom. The motor that operates a ZOOM LENS.

practical. Describing any object, such as a table lamp, that is operable, in contrast to those which are present only for visual effect.

pratfall. A fall that lands the performer on his buttocks, originally made popular by vaudeville comedians.

Praxinoscope. An early horizontal mirrored drum in which different images were located around its inner rim so that, when it revolved, it gave the illusion of images moving in sequence.

preamplifier, pre-amp. Electronic equipment used to amplify sound in a circuit with low voltage signals.

prebreak. To further weaken a piece of BREAKAWAY property by making partial cuts or notches at strategic points.

preemptible. Referring to commercial time sold at a reduced rate that can later be raised to the full rate if the advertiser agrees to it.

preemption. 1. The taking over of schedule network programming time by a local affiliate station, in order to present special programming, usually for live local news events of unusual interest to the community. **2.** The network's cancellation of scheduled programs to air events of national or international consequence.

preflash, prefog. The exposure of film to a little light before its use in order to make it more sensitive to light and reduce contrast.

premiere. The first exhibition of a film to the public; it is often restricted to a single theater, and is a gala event attended by its stars, director, and other filmdom notables.

premium rate. An additional charge made by a station for placing a commercial in a specific desirable time slot.

premium television. Any television transmission system for which a fee is paid by home owners of television sets: e.g., CABLE TELEVISION services.

premix. To combine sound tracks for preliminary audio MIXING.

preproduction. Referring to all activity connected with a film that takes place prior to production.

prequel. The reverse of *sequel;* a new film about the lives of the characters before the story of the first film that portrayed them: e.g., *The Other Side of the Forest* is a prequel to *The Little Foxes,* which was made first and in which the characters are older than in the second film.

prerecord. To record sound that will be played later in the production: e.g., the voice of a dancing/singing star who during the shooting of a strenuous dance routine merely mouths the words.

prescore. To compose and record the sound or music track prior to filming.

preservative. Any preparation, such as lacquer or wax, applied to film in order to protect it.

pressure pad. The pad that holds the tape against the record or playback heads.

pressure plate. A mechanical component that exerts pressure at the GATE of the camera or projector in order to hold the film rigidly in place.

presynchronization. The prerecording of voice tracks in order to synchronize the lip movements in animation work.

preview. **1.** The screening of a motion picture before general release to the public, in order to test audience reaction. **2.** The showing in a theater of brief scenes from a forthcoming film as a "coming attraction."

preview print. See SAMPLE PRINT.

prime lens. A lens of a fixed focal length.

prime time. Peak audience time for television, from 7:00 P.M. to 11:00 P.M. A 1971 FCC ruling (the Prime Time Access Rule) permits only three hours of this daily period to be used for network programming in order to encourage local programming.

principal photography. The scenes in which the principal performers are filmed.

print. **1.** A positive copy made from a film negative, which duplicates the original tonal values and colors. **2.** A positive copy of the complete film. **3.** A TAKE that is to be used in the completed film unless deleted in editing.

printer. An optical duplicating machine that exposes positive film print to light through a negative image, or vice versa, to produce a latent image on the unexposed film.

printer fade. **1.** A visual FADE-OUT (or FADE-IN) made in a PRINTER by the slow reduction of light over the necessary number of frames. **2.** The PRINTER mechanism that controls the decrease of light for the desired number of FRAMES.

printer head. A device in a PRINTER used to make positive prints from negatives.

printer light. The source of illumination in a PRINTER, housed in the PRINTER HEAD.

print film, print film stock. Film designed to carry positive images and sound tracks to be projected on a screen.

printing. Film duplication, often including the addition of a sound track, color adjustments, and special optical effects.

printing light. The calibrated quantity of light used to print a particular scene.

printing negative. The negative print of an optical sound track on a positive picture print, used to produce better sound quality.

printing sync. The synchronization of picture and sound track, which allows for a pull-up track delay.

printing wind, A-wind. The rolling of print film so that the print emulsion and the printing film emulsion will be in contact during printing on CONTACT PRINTERS, or facing each other on OPTICAL PRINTERS; and that the sprocket holes will be in position for correct projection of the final print.

print it. A director's instruction meaning that a take is to be workprinted; the instruction is noted on cameraman's log and on the script itself.

print pitch. See PITCH.

print stock. See PRINT FILM; WORKPRINT (def. 1).

print-through. The undesirable transfer of sound caused by excessive magnetism from one audiotape layer to the next.

prism block. A color-separating optical unit.

prism intermittent. See PRISM SHUTTER.

prism lens. An optical device that produces multiple images within the camera.

prism shutter. A device used on film viewers, editing machines, and certain high-speed cameras that consists of a rotating prism of four or more sides through which the VIEWER light passes as the film is drawn through it.

proc amp (video *proc*essing *amp*lifier). An electronic device used to alter the characteristics of a video signal.

process. To develop and fix exposed film.

process body. The body of any car or vehicle that is shot in front of a PROCESS SCREEN in a studio.

process camera. A camera designed for use in creating SPECIAL EFFECTS, as in MATTE and BIPACK printing, in contrast to cameras used for live-action shots.

process cinematography. Motion-picture photography that uses PROCESS CAMERAS and/or other SPECIAL-EFFECTS cameras.

processing. Developing, fixing, washing, drying, and printing negative motion-picture film.

process photography. Still or motion-picture photography of backgrounds used in BACK PROJECTION shots.

process plate. A positive LANTERN SLIDE that contains a background image to be used for *rear-screen* projection. See BACK PROJECTION.

process projection. Any *rear-screen* projection used to provide a background for action scenes. See BACK PROJECTION.

process projector. A projector that throws moving backgrounds on a translucent screen in front of which actors are photographed during performance.

process screen. A REAR PROJECTION screen on which filmed action takes place behind performers in the foreground. See BACK PROJECTION.

process shot. 1. An action shot made before a BACK-PROJECTION screen on which there are still or moving images. **2.** A combination of film images that appears to have been shot by the same camera.

producer. 1. A person who usually functions as the head of a film production on the artistic level, who coordinates and supervises all facets of the production. **2.** The person who prepares a program for broadcast and is responsible for its economic success.

Producers Guild of America. An organization of motion picture and television producers.

production. 1. The overall process involved in making a film. **2.** The preparation of a program or commercial for broadcast.

production assistant. The principal assistant to the PRODUCER.

production breakdown. The manner in which the film script is divided into scenes or shots prior to shooting, so that the shooting schedule can be arranged in the most efficient manner.

production camera. Any camera used to photograph live action, in contrast to a PROCESS CAMERA or ANIMATION CAMERA.

Production Code Office. The organization that enforces the production code of the Motion Picture Association of America.

production credits. The names of production technicians and other personnel that are shown in the titles before or after a film is run in a theater.

production designer. The person who conceives and plans the overall appearance of a motion picture, creating a personalized environment that will enhance the mood of the film; these concepts are then executed by the ART DIRECTOR.

production house. The facility in which film or videotape commercials are prepared.

production manager. The executive who supervises and coordinates all business and technical arrangements of a film production.

production number. A special number given a film production, used in all financial arrangements.

production personnel. All persons who work on the production of a film from WRITER and DIRECTOR to SCRIPT GIRL and BEST BOY, from GAFFER to CASTING DIRECTOR, etc.

production report. A detailed form that lists all performers and crew members used on a specific day, their work time, the amount of footage shot, and other information useful to the producer in determining the progress of the production.

production schedule. See SHOOTING SCHEDULE.

production script. See SHOOTING SCRIPT.

production still, production still photo. A STILL photograph of a scene from a film, used for promotional purposes.

production unit. A self-contained group consisting of director, cameraman and crew, sound technicians, actors, etc., employed to work on a single film production.

product protection. The minimum time interval between competing commercials, designated by the station.

profile. A demographic breakdown of television audience by sex, age, education, and economic levels.

program. 1. Any television presentation—drama, situation comedy, game show, newscast, etc. **2.** Coded computer-processing instructions.

program film. A short dramatic film, only about four or five reels in length, popular from about 1914 to 1920.

projection. 1. The running of film through a PROJECTOR, which enlarges the images when they appear on the screen. **2.** The practice of television newscasters and commentators of predicting the early results of national elections based on polling data and analyses of voting returns from sample precincts.

projection booth, projection box. The room where PROJECTORS are housed and operated, usually at the rear of the theater and above the audience.

projectionist. The technician who operates PROJECTORS and usually is responsible for their maintenance.

projection leader. A short length of film at the beginning of a reel, which enables PROJECTIONISTS to make rapid and smooth changeovers from one reel to the next during the showing of a long film.

projection speed. The rate at which a film is projected, usually twenty-four frames per second for sound film, eighteen frames per second for silent film.

projection synchronism. The correlation of picture and audio on a PRINT so they will be in perfect SYNC.

projection television. Big-screen home television projection systems of three basic kinds: self-contained cabinets that project from the rear on a translucent screen, single-piece front projectors, and two-piece systems in which the projector is aimed at a separate screen or a wall.

projector. A machine that passes a high-intensity light beam through motion-picture film, as it unwinds from the reel, onto a reflective screen; may simultaneously project the synchronized sound track.

promo, promotional announcement. The broadcast announcement of upcoming programs from a network or station.

prompter. 1. A person who CUES the performer during action shots. **2.** A cylindrical device on which the script, printed in large letters, is rolled up before the performer when he/she is on camera.

property. 1. Any literary work for which the screen rights have been bought by a producer or studio. **2.** PROPS.

property truck, prop truck. The truck that hauls PROPS from storage to the studio or location.

prop list, prop plot. A list of PROPS needed in the production.

propmaker. A craftsman, usually a carpenter, who constructs special PROPS needed in a specific scene or scenes.

prop man. A crew member who is responsible for providing the necessary PROPS.

props. 1. SET articles that are used or touched by actors in a scene; furnishing and objects that are not costumes or part of the scenic structures. **2.** "Props" is also the tag for the crew member who is in charge of supplying the props on the set.

protagonist. The actor who plays the lead in a film.

protection. A clear duplicate film made to be used in the event of some damage to the master film.

protection master. **1.** A film duplicate from which prints can be made as insurance against the deterioration or loss of the original. **2.** A duplicate copy of a sound recording.

protection shot. A shot that can be used in the event the continuity between two shots is erratic.

proxar. A supplemental lens element used in some close-up shots to shorten the focal length.

PSA (public-service announcement). A broadcast announcement made in the public interest, for which the time is donated by the station.

PSSC (Public Service Satellite Consortium). A nonprofit organization created (March 1975) to use NASA SATELLITE transmission for public service.

public access, public-access programming. Referring to cable channels that are offered to individuals and local organizations and institutions for broadcast purposes at minimal fees, in an attempt to let citizens use the powerful medium of television for their own purposes. Subjects range from original comedy to informative reports on community affairs; the only restrictions on subject matter involve the use of commercial advertising, copyright infringement, obscenity, slander, and invasion of privacy.

public-address system. See PA.

Public Broadcasting System. See PBS.

public domain. See P.D.

public-service announcement. See PSA.

publicity still. A photograph made for publicity or advertising purposes of a scene from a film (or some other aspect of the production in progress) before it is released.

Public Service Satellite Consortium. See PSSC.

puff. **1.** Exaggerated praise of a film written for publicity purposes. **2.** A glowing film review by a critic.

pull back. To move the camera away from the subject or action.

pull down. A camera/projector mechanism that advances the film into the GATE frame-by-frame, by means of a CLAW, for exposure or projection.

pull focus. See FOCUS PULL.

pull negative. To match the original negative film to the edited WORKPRINT.

pull up. A LOOP of film used to reduce jerking and to maintain a smooth flow as film passes through the picture GATE over the SOUND HEAD.

punch. **1.** A device used to punch a CUE MARK in a film leader in order to designate when editorial or printing synchronization should begin. **2.** A device used to remove splicing sounds in prints made from an OPTICAL SOUND negative **3.** A device used in making holes in the edge of an animation sheet to fit the pegs on the ANIMATION STAND.

punch line. See TAG LINE.

pup. A small 500-watt spotlight.

puppet animation. The animation of puppets, usually requiring several heads with various expressions and body appendages with several different positions.

push. See FORCE PROCESS.

push in. To move the camera toward the subject or action.

push-off, push-over wipe. An optical effect in which one image appears to be pushed vertically off the screen by another.

Q

quad, quadruplex. A four-unit videotape recorder that rotates at approximately 14,400 rpms at right angles to the transported two-inch tape; delivers video information in successive vertical (or nearly vertical) stripes.

quadlite. A LUMINAIRE unit that contains four 500-watt floodlights.

quarter-inch tape. Standard-width magnetic tape used in tape cartridges and on reel-to-reel recorders.

quarter-load. A powder load in revolvers used in a film, noisy but harmless.

quartz lamp, quartz iodine lamp, Q-I. A TUNGSTEN-HALOGEN lamp containing iodine gas which increases the life of the filament and helps maintain its color temperature.

quartz light. Any LUMINAIRE that uses TUNGSTEN-HALOGEN lamps.

quick cuts. Instantaneous transition shots made without DISSOLVES, so that shots follow each other in rapid succession.

quick save. The obvious manipulation of a film plot so that problems are solved much sooner or more easily than anticipated by the audience.

quick study. A performer who memorizes dialog quickly.

quonking. Extraneous sounds picked up by a microphone.

R

rabbit ears. The V-shaped antenna on some home television sets.

raceway. A recessed channel for cable television.

rack. 1. A frame on which instruments or equipment are mounted. **2.** A roller frame in a film-processing machine. **3.** To pivot a camera LENS TURRET. **4.** To mount reels and thread film into the projection path in order to put the film in frame.

rack focus. See FOCUS PULL.

rackover. A device that enables a VIEWFINDER to be moved to the position behind the lens that is usually occupied by the film.

radio frequency. See RF.

rain cluster. A group of sprinkler heads used to create "rain" on a set.

rain hat. A device used to protect a microphone from rain during outdoor shooting.

rain standard. A pole on which RAIN CLUSTERS are placed.

random access. The capability for easy retrieval of stored electronic information.

random digital dialing. See RDD.

Rank-Cintel Flying Spot Scanner. A machine used with interlocked magnetic sound film to transfer motion-picture film to tape in the production of video cassettes to be used on home television recorders.

rasters. A grid of vertical opaque and transparent slats which separate the left- and right-eye images in stereoscopic (3-D) projection so that special glasses are not required by the audience.

rate card. See CARD RATE.

rateholder. A brief broadcast commercial, aired to maintain the sponsor's weekly schedule continuity and discount allowance.
rate of development. The speed at which a photographic image emerges on film when it is being developed.
rating service. Any research group that conducts audience surveys in order to determine television viewing habits.
rating system. A system adopted by the Code and Rating Administration of the rating board of the Motion Picture Association of America to rate motion pictures according to their viewing suitability for adults and persons under seventeen. See G-RATED, PG-RATED, R-RATED, and X-RATED.
ratio. See EDITING RATIO.
raw stock. Unexposed sensitized negative film in standard lengths, also VIRGIN videotape.
RDD, random digital dialing. A telephone audience survey technique in which households are picked at random.
reach. The number of times during a specified period that viewers can see a television program or a commerical.
reaction shot. A camera shot that cuts to the performer's face to register his/her emotional response to something that has just taken place; usually made as a CLOSE-UP SHOT.
Reader. Studio employee who reads submitted manuscripts to determine if they are potential film material.
reader. See SOUND READER.
reading. 1. The first run-through of a script, without direction, by actors. **2.** The amount of light present according to the reading on a light meter.
readout. The retrieval of electronically stored information.
read-out lines. The lines on lenses and light meters that lead from one scale to another.
read-through. See READING (def. 1).
realism. All elements of a film production, such as costumes, scenery and dialog, which are intended to convey to the audience a sense of reality.
rear projection. See BACK PROJECTION.
rear-projection unit (RP unit). A projector that is used for BACK PROJECTION.
rear-screen projection. See BACK PROJECTION.
recall interview. A telephone audience survey made in order to research the recent television viewing habits of home audiences.
receiver. Combined electronic components that provide visual or audio access to a broadcast.
reciprocating reflex mirror. A mirror that is set at 45 degrees to the lens axis between the lens and the film, used to control light from the lens.
reciprocity law. The longer the film is exposed the greater the density of the developed image; the less it is exposed, the less the density.

record. 1. To store electromagnetic signals for future retrieval. **2.** To make an impression of sound on film, tape, or disk so that it can be reproduced.

recorder. Any device on which sound can be recorded: audiotape, magnetic film, motion-picture film, disks, or wire.

recording. Describing any system used to preserve sound.

recording studio. A soundproof room in which recordings are made.

recordist. A technician who controls the recording equipment.

redressing. To refurnish a set exactly as it was in previous shots.

reduction print. A positive 16mm PRINT made from a 35mm MASTER POSITIVE, usually with some loss in detail and quality.

reel. A flanged metal or plastic spool on which film or tape is wound. A reel usually holds 1,000 feet of 35mm film, 400 feet of 16mm film; a two-inch videotape reel holds about 4,800 feet, a quarter-inch audiotape reel holds about 2,400 feet on a ten-inch reel, 1,200 feet on a seven-inch reel, and 600 feet on a five-inch reel. The 35mm and 16mm reels run approximately eleven minutes each.

reel-to-reel. See OPEN REEL.

re-establishing shot. A camera shot, usually made at the end of a sequence, that shows another angle of the original ESTABLISHING SHOT; may be used to show the passage of time, to indicate a change during the time span of the sequence, or to remind the audience of the context of the closer shots.

re-exposure. The second exposure of a REVERSAL positive film.

reflected light meter. A meter capable of reading light that is reflected from any object in the action area.

reflection. Any light that strikes a surface and is not absorbed by it but "bounces" off to create a secondary source of light.

reflector. 1. Any object with high reflective characteristics that is used to reflect light. **2.** A glass or metal plate in the back of an incandescent lamp which redirects light onto actors or some part of a scene.

reflector lamp. A lamp that contains a REFLECTOR.

reflex, reflex camera. A camera that contains an optical mirror system reflecting the light from the lens into a VIEWFINDER.

reflex focusing. Focusing on the action area by using a camera that contains a reflex system.

reflex shutter. A mirrored shutter that is set at 45 degrees to the lens axis and reflects light into the VIEWFINDER as it passes through the lens.

reflex viewfinder. A VIEWFINDER into which light is reflected, as it passes through the lens, from a partial mirror, a mirrored shutter, or a reciprocating mirror.

regional. Describing network programs fed to television stations within a limited geographical area, usually sponsored by companies whose products are not distributed nationally.

registration. The proper alignment of film in a fixed position so that it will be the same on the artist's table and on the ANIMATION BOARD.

registration pegs. Pegs on the PLATEN used to prepare artwork; adjusted on the animation board so that the framing used on the platen will be exactly the same on the ANIMATION BOARD.

registration pins. See PILOT PINS.

regular 8 film. See CINÉ 8 FILM.

reissue. 1. To return a film to circulation after it has been withdrawn from exhibition. **2.** The film itself.

relational editing. See ASSOCIATE EDITING.

release. 1. The initial exhibition of a film to the public. **2.** News bulletin or information made available to the public.

release negative. A duplicate negative from which RELEASE PRINTS are made.

release print. A final COMPOSITE print of a motion picture, ready for general distribution and exhibition.

relief. Referring to a shot or sequence inserted into a film to reduce the audience tension following scenes of horror, fear, or trouble.

remake. A contemporary version of an old motion picture.

remote. A local broadcast made outside the studio.

renewal. Referring to a television station license reissued by the Federal Communications Commission.

rental film. A film on which a fee has been paid to a distributor for its temporary exhibition.

renter. A person who buys the distribution rights to a film from a producer or other distributor, such as a theater owner, for a specified period of time.

repeat. The rebroadcast of a television program or series.

replenishment. The process of gradually adding fresh photochemicals to a film bath.

reprint. To make a positive film print from a negative.

reprise shots. Camera shots that will be repeated later in a film; e.g., a shot of a couple in a boat just before it overturns will be shown again to remind the audience of the position of each person at that precise moment.

reproduction. Recorded signals transformed into audible sound.

rerecord. To make an audio recording from one or more records; to combine several component tracks (dialog, music, and sound effects) on a single master track.

rerelease. 1. A film shown after it has been removed from exhibition for a period of time. **2.** RERUN.

rerun. 1. A television series that is shown after its initial run. **2.** To show the series again.

residuals. 1. Earnings by performers and others from programs repeated after the initial showing. **2.** Any earnings, usually stipulated in contracts, by performers in addition to their salaries.

resolution. The solving or working out of the protagonists' problems in the climax of a film.

resolver. A device that controls the speed of a magnetic film recorder or a tape playback machine.

resolving power. The ability of a lens or emulsion to make distinguishable the details of an image.

resumé. A written summation of a performer's past employment record and qualifications.

retail rate. A reduced broadcast advertising time rate offered to local merchants, as compared with rates for advertisers who advertise on a national scale.

retake. A shot made of action after a first shot has been rejected.

reticle, reticle lines. The etched lines on a camera VIEWFINDER that indicate the center of a film frame or the SAFE-ACTION areas of television projection.

reticulation. The wrinkling of film EMULSION.

retrieval. The recovery of stored magnetic information.

retrospective. The showing of several films by a particular director or featuring a particular star over a period of several days.

reveal. To pull the camera back from the action area in order to include additional picture material in the frame.

reversal, reversal film. An original film processed to produce a POSITIVE IMAGE, eliminating the intermediate and printing steps.

reversal intermediate. The reversal of SECOND GENERATION DUPLICATE to make it the same type (negative or positive) as the original.

reversal original. A REVERSAL film prepared for exposure in a camera.

reversal process, reversal positive process. The procedure used in developing positives from film exposed in a camera or printed from a positive.

reversal print. A reversal original print that is put through the REVERSAL POSITIVE process. See REVERSAL PROCESS.

reverse action. Shots or prints of film action that goes backward, used for SPECIAL EFFECTS.

reverse angle. A 180-degree change in a camera angle from the preceding shot. See ROUNDY ROUND.

reverse motion. See REVERSE ACTION.

reverse printing. Printing film and print stock run through an OPTICAL PRINTER in opposite directions so that the action will seem to move backward during print projection.

rewind. A high-speed return of tape or film from the take-up reel back to the feed reel.

rewinder. A support on which is mounted a pair of geared hand-cranked devices to wind film from one reel to another.

rewrite. **1.** To revise a film script. **2.** The revision itself.

rf (radio frequency). The waves that transmit video and/or audio electronic signals.

rf modulator. A portable videotape recorder device used to feed recorder playback signals into a local television receiver channel.

rf patter. Picture distortion caused by high-frequency interference.

RGB. Television's basic colors: *r*ed orange, *g*reen, and *b*lue violet.

rheostat. A wire coil used in voltage control.

rhubarb. The murmuring sound made by a crowd in a film shot or sequence.

rhythm. The tempo of a film as created by the length of the shots and the transitions from one to another.

ribbon, ribbon microphone. A directional microphone in which a light metallic ribbon is suspended in a strong electromagnetic field, making it ultra-sensitive to sound.

ride focus. To adjust the lens focus, by use of the focusing rings, in order to keep an action shot sharp and clear.

ride gain, ride the spot, ride the needle. To monitor and control the amplitude of audio signals, either in recording or in transmission.

rifle. See SHOTGUN.

rifle spot, rifle spotlight. A long ELIPSOIDAL spotlight.

rig. To set up equipment, such as flats and lights.

ringing. The dark outlines that appear around televised images.

rip and read. To read news items on the air as they are taken directly from a news-service teletype machine.

ripple. An optical effect that produces wavy or smeared images during film DISSOLVES.

riser. A low platform or table used on the set to support actors or equipment. See APPLE BOX; PANCAKE.

roadblocking. Showing the same broadcast commercial on all local channels in the same time period.

road show. A film exhibited only in major theaters, for which a higher admission price than usual is charged.

rock. To draw tape back and forth across a playback HEAD in order to find specific material.

role. The character played by a performer in a film.

roll. **1.** A length of raw film stock on a spool or core, usually 1,000 feet of 35mm or 400 feet of 16mm film. **2.** The director's verbal command to start film or tape rolling. **3.** The rotation of a camera around its lens axis. **4.** Undesirable vertical movement in a picture.

roll camera. The director's command to the cameraman to start the camera.

roll 'em. The director's command to start the cameras and recorders.

rollers. The pulleys through which film is threaded in the processing equipment.

roll-film motion-picture camera. A motion-picture camera in which rolls of film, rather than cartridges, are used.

rolling title. A moving title that rolls up from the bottom of the screen. See ROLL-UP TITLES.

roll-off. The gradual lessening of high and low sound frequencies.

roll sound. Verbal instruction given to sound technician to start the sound recorder.

roll tape. Instruction to begin showing a segment of tape.

roll-up titles. Titles printed or painted vertically on a strip of flexible background material and rolled on a revolving cylinder so that they appear to crawl up the screen during projection.

roman à clef. A film about famous contemporary figures who are disguised by fictitious names. (From the French: "novel with a key.")

room noise, room sound, room tone. Ambient noise in a room recorded to be used as background sound to fill blanks that occur in the tape after the dialog has been edited.

ROS (run of schedule). The schedule of broadcast advertising over which the local station has control; usually offered at lower than full rates but may be PREEMPTIBLE.

rotary movement. Referring to spinning images produced on film by use of the optical-spin attachment of an OPTICAL PRINTER.

rotary shutter, rotating shutter. A revolving camera or projector shutter that interrupts the flow of light as it passes through; used when filming against a PROCESS SCREEN.

rotoscope. 1. A device with a mirror or prism attachment used to project light through a camera lens in order to accurately position artwork as it moves from frame to frame. **2.** A device that projects single frames to be photographed again or to be traced by hand in the production of artwork or animation detail.

rough cut. 1. The initial editing stage in which shots and sequences are shown in correct context but without the careful and intensive timing and editing necessary for the RELEASE PRINT. **2.** In television, the rough cut is also made to conform to the approximate broadcast time period.

rough-light. Referring to the overall foundation lighting of the set and the action area.

roundy round. To turn the camera 180 degrees from a subject to focus on a subject in the opposite direction.

R-rated. The label given by the Code and Rating Administration rating board of the Motion Picture Association of America to motion pictures that are considered unsuitable for persons under seventeen unless accompanied by an adult.

RTNDA. The organization of the professional members of the Radio/Televison News Director's Association.

rub-off animation. See SCRATCH-OFF.

rumble pot. A receptacle of boiling water on which dry ice is floated, used to create the illusion of fog.

run. 1. To schedule or transmit a television program. **2.** The length of time, in days or weeks, during which a motion picture is shown at a specific theater.

runaway, runaway production. 1. A production made on an American location away from Hollywood or in a foreign country for the purpose of reducing production expenses. **2.** A production designed to bypass the use of union performers and crew members.

rundown. The order in which program events are to be broadcast.

running gag. The repetition of a humorous bit throughout a film or television program.

running part. A continuous role in a daily or weekly television series.

running shot. A shot in which the camera, usually hand-held, keeps pace with the performer, who is in motion.

running time. 1. The length of time a film will run when projected at normal speed. **2.** The length of a broadcast in actual on-air program minutes.

run of schedule. See ROS.

run-through. A cast rehearsal monitored by cameras and recorders that are not operating.

rushes. See DAILIES.

S

safe-action area, safe area. The picture area within the frame of film that will not be eliminated during transmission; about 90 percent of the screen, measured from its center.

safebase. A slow-burning film base such as cellulose triacetate.

safelight. Colored light used in darkroom illumination that prevents adverse effects on the photographic emulsion while permitting workers to see what they are doing.

safety. See SAFE-ACTION AREA.

safety film. Film on which a SAFEBASE is used.

SAG (Screen Actors Guild). The trade union of performers in motion pictures.

saga. An epic-like drama that usually covers several generations in a family: e.g., *Roots.*

salamander. A stage heater used during cold weather.

sample print. A COMPOSITE print that has not been approved for release.

sandbag. A canvas bag filled with sand used to anchor scenery, set pieces, light stands, etc.

Satcom III-R. An RCA SATELLITE that carries various cable service networks including Home Box Office (pay-cable), the Black Entertainment Network, the Spanish-language Galavision, etc. Satcom I, launched by RCA in 1975, was the first television satellite.

satellite. 1. Any orbiting space station used to relay distant transmission signals (in one-quarter of a second). A DISH on the ground sends and receives signals to and from the satellite. **2.** A separate broadcast facility that retransmits material of a nearby station to improve and increase its local coverage. **3.** See AERIAL, ANTENNA.

Satellite News Channel. A twenty-four-hour all-news network satellite scheduled to be launched in 1982 by ABC and Group W Satellite Communications, which will own it jointly.

Saticon. A sensitive television camera pickup tube with a target surface coating of selenium arsenic tellurium.

satire. A film or television program in which certain aspects of a particular culture or individuals are ridiculed.

saturation. 1. The persistent exposure of a commercial message to home viewers. **2.** The intensity of picture color having a purity of hue, value, and chroma.

save it. See PRINT IT.

save the arcs, save the lights. A command, usually from the director, to turn off the set lights.

SAWA (Screen Advertising World Association). A trade group that promotes advertising in motion-picture theaters.

scale. Graduated minimum union rates paid to television performers or program guests.

scallop. A wavy distortion in a television picture.

scan. The horizontal electron beam sweep across the television camera target or picture tube in one-fifteenth of a millisecond; a full vertical scan is one-sixtieth of a second.

scan line. See SCANNING LINE.

scanning. 1. The action of an electron beam as it traces a pattern over the TARGET portion of the camera pickup tube; the conversion of optical sound film to electronic signals. **2.** Directing a narrow light beam or other electromagnetic radiation over an entire area.

scanning line. A single horizontal path made across a television picture tube by an electronic beam.

scatter plan. A television broadcast advertising schedule arranged to achieve maximum coverage of an audience chosen at random.

scenario. 1. An outline of a film script that describes the action in sequence and includes brief descriptions of scenes and characters. **2.** A term occasionally used to describe the complete film script.

scenarist. A person who writes screenplays, outlines, or treatments.

scene. 1. A setting for a particular shot or series of shots. **2.** A series of interrelated shots made in continuous action.

scene contrast. The arranging of color tones in an action area so that they can be changed quickly from light to dark.

scene dock, scenery dock. An area in which painted BACKDROPS and FLATS are stored.

scenery. 1. Outdoor elements (trees, mountains, permanent buildings, etc.)

used as backgrounds. **2.** Artificial backgrounds used on the set or location.

SCG (Screen Cartoonists Guild). The trade union for artists and others who work on animated films.

schedule. The dates and times of broadcast commercials according to commitments made by advertisers.

schematic. Referring to the wiring diagram for electronic equipment.

schlieren lens. An optical device used for video projection.

schtick. A routine used by an actor, usually a comedian.

Schufftan process. A SPECIAL-EFFECTS process in which mirrored backgrounds are photographed with action seen through a mirror from which portions of the silver have been eliminated. Used to mask an area in which other figures will appear on the film.

science-fiction film, sci-fi film. A film based on fanciful scientific phenomena and explorations that usually involve dangerous adventures.

sciopticon. A device that projects moving slides.

scoop. See BASHER.

scope. A CATHODE-RAY TUBE device used in the analysis of visual electronic signals.

score. **1.** To compose the musical accompaniment for a film. **2.** The music itself. **3.** The music that accompanies a broadcast program.

Scotchlite. The trade name for a highly reflective sheet of background material.

Scotchlite process. A FRONT PROJECTION that uses a Scotchlite background.

scouting, scouting locations. Preproduction searching for suitable outdoor sites to be used for LOCATION during filming.

scramble. **1.** To encode an electronic signal transmission; *unscramble* means to decode the transmission.

scraper. A sharp SPLICER device used to remove film EMULSION before applying cement.

scratch. **1.** An abrasive mark made on film; white scratches on negative film, black scratches on positive film. **2.** Mechanical damage to the oxide coating of videotapes.

scratch-off, scratch-off animation. Artwork ANIMATION, made with water-soluble paint and shot upside down, in which areas of the artwork are removed (rubbed off) between exposures and the processed film is turned end-for-end to make the removed image reappear.

scratch paint, scratched print. **1.** A print, usually black and white, made from an edited WORKPRINT, used during post-production work for dubbing and mixing. See DUB; MIX. **2.** A small POSITIVE stock shot that is deliberately damaged to prevent its being used for projection or illicit duplication; the original negative is kept for future use.

scratch track. A sound recording made to be used only for editing purposes.

screen. 1. A reflective sheet of material on which motion pictures or backgrounds are projected. **2.** To project a motion picture. **3.** Generally, motion pictures themselves.

Screen Actors Guild. See SAG.

Screen Advertising World Association. See SAWA.

screen brightness. The degree of luminance, measured in candles per square foot, given off by a screen.

Screen Cartoonists Guild. See SCG.

screen credits. 1. The titles that list the names of the cast and other creative members of a film, shown at the beginning or end of a film. **2.** A list of films in which an actor or creative artist has worked.

screen direction. Direction of actors and action within the frames of a film.

Screen Extras Guild. See SEG.

screening. The presentation of a film to a selected audience.

screening room. A facility in which films are shown to small nonpaying groups; many screening rooms are built into the private residences of filmdom's elite.

screenplay. A script written for a motion picture, the basis of its dramatization and production.

screen ratio. See ASPECT RATIO.

screen test. An audition by actors in which cameras and sound equipment are used.

screenwriter. A person who writes scripts, treatments, stories, or outlines for motion pictures.

screwball comedy. Motion pictures, first popularized in the 1930s, in which characters are involved in zany and improbable situations.

scribe. A metal stylus used by film editors to jot information on the edge of film.

scrim. Gauze or netting (occasionally metal) light diffusers used to reduce the intensity of natural or artificial light.

script. Material written for a film, which includes not only the plot outline but also brief descriptions of characters and settings, and complete dialog, narration, and limited action and sound effects. Scripts undergo many revisions before and during production; the final version is called the SHOOTING SCRIPT.

script clerk, script girl, script person. See CONTINUITY CLERK.

scriptwriter. 1. See SCREENWRITER. **2.** In television, a professional writer of broadcast material.

scrub. To eliminate material.

SE. See SOUND EFFECTS.

secondary service. See SKYWAVE.

second feature. The second film, usually considered the less important film, in a double-feature presentation.

second generation duplicate, second generation dupe. A duplicate film made from an original.

second unit. A backup crew on a film production, usually working on location filming.

Section 315. The EQUAL TIME section of the Federal Communications Commission Act (1934).

Section 326. The Free Speech section of the Federal Communications Commission Act (1934) in which FCC censorship is prohibited.

SEG (Screen Extras Guild). The film extra's trade union.

segue. 1. To dissolve one piece of music or audio element into another. **2.** The end of a piece of music immediately followed by the beginning of another. **3.** A self-contained piece of film action that usually begins and ends with a FADE or it can be ended with a DISSOLVE or CUT.

selected take. The final approved version of a taped or filmed shot or scene.

selective focus. Shooting in sharp focus only a section of the action area.

selectivity. Receiver discrimination between two adjacent broadcast signals.

self-blimped. Referring to a special camera designed to produce a low noise level during operation so that camera noise will not have to be blimped out in editing.

self-matting. An optical process in which color MATTES are used to eliminate rotoscoping. (See ROTOSCOPE.)

selsyn, selsyn motor. An electric motor that automatically synchronizes projection interlock systems, remote control mechanisms, etc.

semiconductor. Material capable of electron transfer when tiny voltages are applied.

senior, senior spot. A 5,000-watt spotlight.

sensitivity. The degree to which a film EMULSION responds to light.

sensitometer. A device used to expose film with accurately measured EMULSION speed.

sensitometric strip, senso strip. A filmstrip that has been exposed in a SENSITOMETER to evaluate its photographic response or processing conditions.

sensitometry. The science of determining the exposure and development characteristics of photographic materials.

separation. 1. PRODUCTION PROTECTION. **2.** The breakdown of color values into the primary colors. **3.** The decibel ratio between speaker channels.

separation positives. See COLOR SEPARATION.

seque. See CROSS FADE.

sequel. A film production that continues the story of characters who appeared in a previous film or films: e.g., the James Bond movies.

sequence. 1. A series of interrelated shots in which a single subject is shown. **2.** The sequential order in which the shots are shown.

sequence shot. A long shot in which both actors and camera move around; made in order to eliminate separate setups for closer shots. Often used when action is too involved to break into separate close or medium shots.

serial. 1. An early form of EPISODIC motion pictures that ends with a CLIFF-HANGER, to be continued in the next episode, which begins the next movie. **2.** A continuing daytime television series (soap opera).

series. 1. A motion-picture genre popular in the 1940s in which the same characters appear in several feature films: e.g., the Andy Hardy series, starring Mickey Rooney. **2.** Situation comedies and other television films in which the same casts appear weekly.

servo motor. A motor system in which the speed control is self-regulating; the output is used to control the input.

set. 1. The indoors area where filming takes place. **2.** In a studio an artificial construction designed to give the illusion of a real location: e.g., a living room or a precinct office. **3.** A television set or receiver.

set-and-light. Order given by director to prepare for shooting.

set decorator. The crew member who is responsible for the construction of sets designed by the SET DESIGNER.

set designer. The person who creates and sketches the sets for a film.

set dresser. The crew member who is responsible for the placement of furnishings on the set.

set dressings. All furnishings on a set.

set light. See BASE LIGHT.

setting. 1. The location and time in which the action in a motion picture takes place: e.g., New York City in the 1880s. **2.** The location as constructed for the film. **3.** The natural location: e.g., forest or open range.

set up. 1. To put a camera and its associated lights into position. **2.** Generally, to put all sound, camera, and lighting equipment needed for a particular scene into position.

7½ ips. The recording speed (inches per second) of audiotapes.

750. A baby spotlight.

70mm film. An early camera film now used chiefly for making reduction duplicates.

sexploitation film. A motion picture in which the emphasis is on explicit sex, probably X-rated; usually of inferior quality as far as acting, directing, lighting, and cinematography are concerned.

SFX. See SOUND EFFECTS.

shader. The video engineer who controls the quality of a television picture for the switcher.

shading. The contrast adjustment in a television picture.

shadow box. A device used to adjust lights to control shadows on the set.

shadow mask. The perforated mask behind the face of the color television picture tube, which separates the basic color (red-green-blue) electron beams.

shallow depth, shallow focus. A limited area in the action field which is in sharp focus, with the surrounding area blurred.

share, share point. The Nielsen-computed percentage of persons who are actually watching a given program, determined by dividing the rating by the number of persons watching: e.g., if the meters indicate that 60 percent of the Nielsen members had their sets on during a particular program and 30 percent of them were actually watching it, the program's share would be 50. See NIELSEN.

shared I.D. The usual station identification added to commercial copy on film, slide, or card.

share the boob. Referring to performers or material that have equal exposure on television.

sharp. Describing **1.** images that are clearly in focus, **2.** a camera lens capable of recording image detail.

shielded cable. An inner signal conductor that is protected from unwanted random signals by grounded metallic braid.

shoot. 1. To photograph action. **2.** To film or tape material to be used in a film.

shooting. All action (by cameras, sound equipment, actors, directors, etc.) involved in the actual filming process as it occurs.

shooting call. Instructions to the film crew about when and where to report for the day's shooting; e.g., "11:00 for 12:00, AOC, Sunset and Vine, having had" means to report at 11:00 o'clock with all necessary equipment to begin shooting at 12:00 o'clock, "ass on curb," at the corner of Sunset and Vine, having had lunch.

shooting date. The day scheduled for filming or taping specific sections of a film.

shooting log. A printed form on which a camera crew member records the type of film, the kind of camera, and the filters used in shooting specific scenes, information about the length of each take, and whether or not the take is to be workprinted.

shooting ratio. The amount of film shot compared with the length of the film that is finally edited.

shooting schedule. A form on which is listed the shots to be made on specific dates, the names of the actors, and lists of the production personnel and the kinds of equipments needed. Shots are listed for expediency and not always made in the sequence in which they will be shown.

shooting script. The final approved script, usually with numbered scenes or shots, to be used by actors and the director during filming.

shooting the playback. Filming actors during the playback of prerecorded sound while they move their lips in synchronization with the sound. This reduces sound recording problems on the finished film.

shooting upside-down. Shooting done with the camera turned upside down, after which the film is turned end-for-end in projection or printing in order to make the action seem to move backward.

shoot on speculation, shoot on spec. To shoot a film without a prior contract with a buyer, hoping to recoup expenditures through the sale of prints.

short. 1. Program material not long enough to fill the time allotted for it. **2.** SHORT CIRCUIT. **3.** SHORT SUBJECT.

short circuit. Contact between two parts (usually exposed) or wires in an electric circuit, often causing functional failure in the equipment.

short end. Unexposed film, too short to be used, near the end of a reel of raw stock.

short focal-length lens, short-focus lens. See WIDE-ANGLE LENS.

short pitch. The standard distance between the leading edges of SPROCKET HOLES in film to be used for original prints.

short skip. A minimal transmission signal reflection (approximately 100 to 1,000 miles).

short subject. A one-or-two-reel film, not over thirty minutes of running time; once popular in theaters but seldom seen today.

shot. A recording of action or images by the motion-picture camera in a brief series of frames, usually 10 to 15 seconds long. Kinds of shots include ESTABLISHING, MEDIUM (see MCV), CLOSE-UP, LONG, DOLLY, PAN, TRAVELING, RUNNING, TRACKING, ZOOM; also, camera angles provide additional varieties of shots such as LOW and HIGH ANGLE and TILT SHOTS.

shot analysis. A written description of each shot in order to study it in relationship to the total film.

shot box. A device, set prior to filming, that controls the ZOOM LENS system in a camera.

shot breakdown. 1. A list of shots to be made in sequence, plus a list of actors, crew members, equipment, and location needed for each one. **2.** A list of the various camera positions required to film specific action shots.

shotgun. A long microphone used for directional sound pickup, capable of picking up sound over a great distance.

shot list. 1. A list of shots in which there is no narration or dialog. **2.** A card listing shots to be made by a television cameraman who is filming live.

shot plot. A drawing, on the plan of a set or location, of camera angles that cover action written in the script in order to indicate the focal length of the lens to be used.

shot sheet. See SHOT LIST.

shoulder brace, shoulder pod. A support for a hand-held camera, which conforms to the contours of the body.

shrinkage. The reduction in film size due to a loss of moisture while drying after processing or from being stored for a long time.

shutter. 1. A rotating device that in a camera protects the film from light at the APERTURE, and in a projector reduces the projection light while the film is moving at the aperture. **2.** An INTENSITY control device on a spotlight.

shutter angle. The degrees in the angle of a shutter opening.

shutter control. A control on a camera which is used to partially close the shutter in order to reduce exposure or to open and close it gradually in order to achieve FADE-IN and FADE-OUT.

shutter opening. See SHUTTER ANGLE.

shutter release. See CABLE RELEASE.

shutters. The vertical slat diffusers attached to a large 10,000-watt spotlight.

shutter speed. The amount of time a SHUTTER is open during the desired exposure of a frame.

SIA (Storage Instantaneous Audimeter). The revised AUDIMETER used in obtaining faster readouts for the NIELSEN ratings.

sibilance. A hissing vocal sound, usually on the letter *s* or a soft *c* followed by a vowel.

sideband. The RF-modulated area above the carrier frequency.

sight gag. A joke that is visual only, or a silent comic bit.

sight line. The line of vision to the screen from a person seated in a commercial theater.

signal. An electrical impulse converted to pictures or sound.

signal-to-noise ratio (S/N). The difference measured in decibels between a fully modulated signal and the extraneous noise that exists in the transmission system itself, where there is no modulation.

signature. Identifiable music associated with a specific product or advertiser.

signer. A person who uses sign language to interpret television programs to hearing-impaired viewers as the programs are being broadcast.

sign-on, sign-off. Station identification that is broadcast at the beginning and end of each day's transmission.

silent, silent film. 1. Film prepared or projected without a sound track. **2.** Any film made before sound-motion pictures (1928); usually accompanied by piano or organ music played by a musician located in the pit below the screen.

silent speed. The exposure rate of film, which meets the requirements of persistence of vision; currently the speed is eighteen frames per second, originally was sixteen frames per second. (During silent films the cameraman or the projectionist could vary the speed by cranking slower for romantic scenes, faster for strenuous action scenes, etc.)

silhouette. The dark outline of a performer, created by pure back lighting.

Silicon-Intensified Target. See SIT.

silk. A light diffuser or diffusing reflector made of gauze or a sheet of white fabric stretched on a frame.

silver halide. A light-sensitive compound used in the production of film EMULSIONS.

silver-oxide battery. A small coin-sized low-power battery cell, usually long-lived but unreliable.

simplex. Temporarily replacing regular aborted broadcasting with another transmitting service.

simulcast. To televise and transmit by radio at the same time.

simultaneous contrast. The alteration of monochromatic tones or color values in relationship to their surrounding: e.g., medium gray appears dark gray against a light gray background.

single broad. A cube-shaped 2,000-watt FILL LIGHT.

single/double-system sound camera. A camera that has a BLIMP and is used for single- or double-system sound filming.

single-frame release. A button, lever, or cable release on a camera, which can be operated to expose single frames.

single-frame shooting, single framing. Using the release mechanism on a camera for each frame of exposed film in order to speed motion when film is projected at normal speeds: e.g., to show the growth of plants or animals as in TIME-LAPSE CINEMATOGRAPHY.

single perf. 16mm film with a SOUND TRACK along one edge and SPROCKET HOLES along the other.

single-rate card. A television station's identical fee for both local and national advertising.

single scrim. A single layer of gauze or thin material (such as netting) used to reduce illumination.

single-shot technique. Making shots individually rather than by the MASTER SCENE technique.

single sprocket. See SINGLE PERF.

single strand. Successive negative sequences optically printed on a single strip of film.

single-system. Referring to a camera that records visuals and sound for developing on the same piece of film. The audio recording may be on a magnetic or optical track.

single-system sound camera. A SELF-BLIMPED camera that records both visuals and sound on one strip of film.

single-system sound recording. Simultaneous recording of sound and the action that accompanies it.

siphoning. Pay television cable transmission of a program that has been seen, or has been available, by free broadcast.

SIT (Silicon-Intensified Target). A television camera pickup tube created for low light levels; has high sensitivity and good resolution.

sitcom. See SITUATION COMEDY.

situation. The relationships and conflicts between characters and events.

situation comedy. A comedy series in which the same characters appear regularly, usually weekly, and are involved in situations that arise from their interrelationships and reactions to the environment.

16mm. Film stock that is 16mm wide with either single or double perforations; adopted as international standard in 1923; has forty frames to the foot, at .6 feet per second at sound speed (twenty-four frames per second).

625-line. The standard scanning rate per frame throughout the Eastern Hemisphere (except Japan) for all television transmission systems; has better picture resolution than the 525-line standard used in the United States.

skew. A zigzag television picture distortion created by irregular horizontal vacuum guide alignment in the videotape recorder.

skin, skin flick. A motion picture that emphasizes sex and nudity; usually a film of inferior quality.

skip frame, skip-frame printing, skip printing. An OPTICAL PRINTING procedure used for SPECIAL EFFECTS in which every second or third frame is skipped in order to give the illusion of speeded-up action and/or to reduce the length of the film.

skipping effect. See STROBING.

sky filter. A lens FILTER that contains color only on the upper half, used to reduce the effect of atmospheric haze.

sky light. Light from the sky exclusive of sunlight.

sky pan. A CYCLORAMA floodlight that produces an extremely wide beam of light. See CYC STRIP.

skywave. The secondary portion of a television broadcast signal that radiates skyward.

slapstick. Describing comic acting that concentrates on violent or exaggerated actions such as PRATFALLS.

slate. **1.** A small blackboard containing identifying information about each TAKE and photographed for several frames before each take; often has a hinged clapper or CLAPSTICKS attached. **2.** A direction, "Slate this take."

slate board. See SLATE.

slate it. See SLATE.

slave. **1.** Any electrical device that can be controlled automatically from a distance. **2.** A recorder that DUBS playbacks from a master track. **3.** The videotape used for such dubbing.

sleaze. Low-budget shock and exploitation films, first produced in the 1960s; they include hard- and soft-core pornography, sci-fi, motorcyclist, and kung fu films, usually ineptly performed, plotted, directed, and produced.

sled. A bracket support that attaches a LUMINAIRE to a wall.

sleeper. A motion picture, usually a low-budget one, which becomes an unexpected popular success.

slice-and-dice film. A genre of horror films showing explicit violence in which the victims are usually young women or teenage girls.

slice of life. A technique used in broadcast advertisements in which (supposedly but not actually) the conversations of real people rather than actors are reproduced.

slide. 1. A drum-mounted transparent photograph projected into a camera chain for TV broadcast. **2.** A transparency intended to be projected by transmitted light.

slide film. A series of 35mm slides shown individually in a cartridge slide projector, sometimes accompanied by a separate synchronized sound track.

slop print. An unedited film print made from an OPTICAL NEGATIVE and used to check mechanical printing errors.

slop-test. The development of a short section of film in portable equipment in order to check for printing defects.

slop-test processing. The processing of short sections of film in portable equipment in order to check for defects.

slot. 1. A groove in the camera box into which FILTERS or MATTES can be inserted. **2.** A time period in which a production is scheduled for recurring appearances.

slow. Describing film EMULSIONS that have less sensitivity to light and are usually less grainy than FAST emulsions.

slow lens. A lens with a maximum aperture of f/2.8 or less.

slow-mo, slow-motion. Referring to videodisc equipment used to achieve SLOW-MOTION, SPEED-UP, FREEZE-FRAME, or REVERSE-ACTION effects; records thirty-nine seconds of time without compression or selective condensation.

slow motion. The illusion of filmed action that is much slower than normal action, achieved by operating the camera at a rapid rate and projecting the film at standard speed; often called high-speed photography.

slow scan. A transmission system used for inanimate subjects.

slow speed. A camera speed that is slower than normal; used to achieve the effect of fast motion during projection at standard speed.

slug. A blank LEADER strip.

slug in. To insert a blank LEADER strip to increase a picture WORKPRINT or sound-track workprint to the necessary length.

smear. A blur on a television screen caused by a moving HOT SPOT or light source.

SMPTE, simpty (Society of Motion Picture and Television Engineers). A standard-setting professional group of engineers in the film and television industries.

SMPTE test film. Any of a series of films from the Society of Motion Picture and Television Engineers used to test equipment against possible malfunction.

SMPTE universal leader. See UNIVERSAL LEADER.

sneak. Slow FADE-IN or FADE-OUT of sound or picture.

sneak preview. The unadvertised presentation of a feature film prior to its general release in order to judge audience reaction.

snoot. A tubular attachment on a spotlight used to reduce the size of its light beam and facilitate pinpointing the beam.

snow. The breakup of a picture with a snowfall effect, caused by weak video signal reception.

snow effects. Pertaining to falling "snow" made on the set, usually with white corn flakes blown by a fan.

soap opera, soap, soaper. A daytime melodramatic series whose name comes from the frequent sponsorship by soap manufacturers on radio and during the early days of television; generally shown daily Monday through Friday. (Weekly series such as *Dallas* and *Knots Landing,* though shown at night, are occasionally referred to as soap operas because they fall within the general format.)

soapdish. A plastic container for an audio cassette.

Society of Motion Picture and Television Engineers. See SMPTE.

sodium light. Illumination of a screen by sodium vapor to provide a yellow light, the one color that cannot be photographed; used to create clear backgrounds for TRAVELING MATTE shots.

sodium process, sodium vapor traveling matte process. A procedure of making a MATTE by photographing action against a yellow screen illuminated by sodium vapor lamps.

sodium thiosulfate. A fixative used in the photograph-developing process.

SOF (sound-on-film). 1. Film footage that is accompanied by sound as the film is being exposed, usually shot by a 16mm single-system camera. **2.** The camera itself.

soft. See SOFT FOCUS.

soft copy. A readout on a CATHODE RAY TUBE.

soft cut. An extremely short DISSOLVE.

soft-edge wipe. A WIPE with blurred edges.

soft focus. 1. An intentional reduction of an image's sharpness by use of an optical device such as a DIFFUSION disk or netting or gauze placed over the lens. **2.** An image (or portion of it) that is out of focus.

soft light. A LUMINAIRE that provides a bright diffused light.

soft porn film. A motion picture in which the emphasis is on sex but in which the sexual acts are simulated rather than actual.

software. 1. Films, audiotapes, and audio disks that are available commercially. **2.** Broadcast program material that is presented on electronic equipment.

solarization. A flare seen in a photographic image due to light having struck the film during development.
solenoid. An electromagnetic switch.
solid state. Referring to a transistorized circuit.
sonic. Describing a sound within audible range, 20–20,000 Hz.
sonic cleaner. See ULTRASONIC CLEANER.
soubrette. 1. The role of a young coquette played by an actress. **2.** The role of a dance-hall singer/dancer around the turn of the century.
sound. 1. Generally, all audio elements in a film. **2.** A SOUND TRACK.
sound advance. The interval on film between a particular frame and the point on the SOUND TRACK that is synchronized with it; sound and picture are reproduced from different points on the projectors.
sound camera. A blimped camera that has special equipment to achieve synchronization with the sound recorder that is used concurrently with it. See BLIMP.
sound crew. Studio personnel responsible for recording sound during filming.
sound dissolve. A brief period when the sound from two SOUND TRACKS overlaps.
sound drum. A flywheel used to maintain the smooth movement of film past the projector SOUND HEAD.
sound editor. The person responsible for assembling, synchronizing, and editing all SOUND TRACKS on a film; he also supplies CUE SHEETS for the SOUND ENGINEER to use in DUBBING.
sound effects. Recorded or live audio effects other than synchronized dialog, music, and narration that create the illusion of reality, added to the master track during mixing.
sound-effects library. A stored and cataloged collection of SOUND EFFECTS on sound tracks, disks, or tapes.
sound-effects man. The technician who is responsible for providing broadcast sound effects.
sound-effects track. See EFFECTS TRACK.
sound engineer. The person responsible for the actual dubbing of SOUND TRACKS, following cue sheets provided by the SOUND EDITOR.
sound head. 1. A device in which sound is picked up from optical or magnetic tracks. **2.** A magnetic recording or erase head.
sound log. A printed form on which a technician records the ROLL (def. 1) and TAKE numbers for sound recordings; a take-by-take record.
sound loop. An endless loop of sound-track film run continually during a MIX (def. 1) as a source of a sound effect where needed; magnetic film and audiotape can also be used.
sound mixer. See MIXER.
sound-on-film. See SOF.
sound perspective. The impression of sound being heard from a distance.

sound reader. A film-editing device, either optical or magnetic, used for the playback of sound tracks.

sound recording. 1. Recording sound on film, disk, or tape. **2.** The recording itself.

sound report. See SOUND LOG.

sound speed. The standard exposure rate of sound film, twenty-four frames per second.

sound stage. A soundproofed studio.

sound stock. Film designed for making motion-picture SOUND TRACKS.

sound stripe. See STRIPE.

sound tape. Thin plastic tape without perforations, coated with iron oxide.

sound track. 1. The audio portion of film or videotape. **2.** Film that contains sound only.

sound-track applicator. A processing machine device used to apply a separate chemical solution to the sound-track area after certain kinds of prints have been partially processed.

sound transfer. See TRANSFER.

sound truck. A vehicle containing sound-recording equipment, including a battery system for power.

sound workprint. A SOUND TRACK used during the editing process.

south. Describing the bottom portion of an ANIMATION chart.

soup. Any solution used in the development of film.

space opera. A science fiction (sci-fi) motion picture, usually one featuring spacecrafts: e.g., *Star Wars*.

spacer. A hub positioned between reels or rewinders used to keep the reels in position as the film moves in or out of a SYNCHRONIZER.

spaghetti Western. A WESTERN, filmed in Italy with an Italian cast except for one or two Americans in starring roles; English dubbed in.

sparks. Nickname for a set electrician.

speaker. A device that transduces electronic signals into sound waves, used to amplify sound. See TRANSDUCER.

spec film. See ON SPECULATION.

special. 1. A major network production usually created to showcase the talent of a single performer: e.g., *An Evening with Fred Astaire*. **2.** A one- or two-hour documentary about some event or problem of current national or international concern.

special effects. 1. Extraordinary effects that normal camera techniques are insufficient to provide and which are usually inserted in the film by the special-effects department after shooting; achieved through the use of traveling MATTES, still mattes, MICROPHOTOGRAPHY, MULTIPLE-IMAGE montages, SPLIT SCREENS, and special printing techniques. **2.** Mechanical techniques used to achieve simulated explosions and fires, the uneven motion of vehicles (boats, trains, stagecoaches), and other manipulated physical visuals. **3.** Graphic elements such as WIPES, INSERTS,

DISSOLVES. **4.** Generally, any device or technique used to create an appearance of reality in a scene where it would be impossible, impractical, or unsafe to use an actual action or effect.

special-effects generator. A unit used in video production which processes several video signals in order to produce a final signal, called the program signal.

special events. Filmed events of general interest, such as the inauguration of the president of the United States or the launching of a spacecraft, which have been scheduled in advance (exclusive of sports events).

specifications, specs. Written instructions for the construction of special equipment or structures.

spectacle film, spectacular. Any feature film, usually a period or historical film, in which everything—the sets, the costumes, the actions, the characters themselves—appears on a grand scale.

specular. The reflection from a performer's eyes or teeth.

specular light. See HARD LIGHT.

speculation film, spec film. See ON SPECULATION.

speed. 1. The light sensitivity of a film emulsion. **2.** The exposure index of film stock. **3.** The largest F-STOP on a camera lens. **4.** The correct operating speed of a camera or projector. **5.** A verbal indication ("Speed!") that the camera or recording equipment has reached its correct synchronous speed.

speed lines. The lines painted behind animated subjects to give the impression of fast motion.

speed up. A hand cue given to a performer to increase his or her tempo.

spherical aberration, spherical distortion. A lens malfunction that creates a curved effect in square images.

spherical optics. Camera or spotlight lenses that have surfaces with various curvatures, used to alter the paths followed by rays of light.

spider box. A small, portable, multi-outlet electrical cable box into which lights and other electrical equipment can be plugged.

spider dolly. A DOLLY equipped with projecting legs to which the wheels are attached.

spill, spill light. Undesirable illumination produced by a scattering of light along the main beam of a light source.

spill-in. A viewing audience from outside a television station's normal area of dominant influence (ADI) or designated market area (DMA).

spill-out. Viewing stations from outside an audience's normal ADI or DMA. See SPILL-IN.

spin. A whirling image produced on an OPTICAL PRINTER or ANIMATION STAND.

spindle. The rotating shaft of a film or tape rewind system.

spirit gum. A sticky substance used to attach false beards and mustaches to an actor's face.

splash light. Undesirable light that falls on a BACK-PROJECTION screen caused by LUMINAIRES directed on the performers in front of it.

splice. 1. To join two pieces of film together by one of several methods including taping, cementing, or butt-welding. (See BUTT SPLICE.) **2.** The joint itself.

splicer. A mechanical device used to accurately join film frames with transparent tape or cement.

splicing block. A grooved device used to hold together the ends of film or audiotape while the SPLICING TAPE is applied.

splicing tape. A pressure-sensitive tape used on the base side of an audiotape splice.

split-field lens, split-focus lens. A camera lens designed to achieve accurate focus at two different planes.

split-focus shot. A shot in which the camera focus is changed from one plane to another.

split screen, split-screen effect. 1. A film frame divided into two or more areas containing separate images, done either in the camera or in an OPTICAL PRINTER (def. 1). **2.** In television the same effect is achieved by combining the output of two cameras focused on two different subjects.

sponsor. 1. The organization for which a film is produced under contractual agreement. **2.** The advertiser for a broadcast commercial.

spool. A flanged device used for holding a roll of film.

spool bank. The rollers on which a FILM LOOP is carried.

spool capacity. The size of a camera film chamber according to the largest daylight-loading film spool that it is capable of accepting.

sports. 1. A special broadcast category, especially athletic events that are commercially sponsored. **2.** A segment of regular news broadcasts.

spot. 1. Any broadcast commercial. **2.** A directional spotlight with a narrow beam that can be adjusted for correct focus by means of a reflector and lens.

spot brightness meter, spot meter. A light meter capable of reading reflected light at a narrow angle of acceptance.

spot news. Unexpected new events broadcast shortly after they have occurred inserted in a regularly scheduled newscast, often by a newscaster who has received no film on the subject.

spotter. An assistant to a sports announcer, usually closer to the field of action than the announcers' booth at the stadium or arena.

spotting. Undesirable blemishes on film that occur during processing.

spotting in. Adding music and/or sound effects to a film in correct context.

spray. To minimize a reflected glare on film with an aerosol MATTE finish.

spray processing. A film-processing technique that passes the film through a chemical spray instead of submerging it in a bath.

spread. 1. The width of a LUMINAIRE beam. **2.** To diffuse a luminaire beam.

spring-drive camera. A camera run by a SPRING-DRIVE MOTOR.
spring-drive motor. A governor-controlled motor run by a hand-wound spring drive.
sprocket. 1. A film roller transport system in which the toothed gear engages the perforations in the film edges to move the film through the various types of mechanisms. **2.** One of the projections on the roller.
sprocket holes. The perforations or holes in the edges of film, punched at regular intervals.
sprocket roller, sprocket wheel. A roller equipped with teeth that engage the perforations of film to move it forward in the various mechanisms.
spud. A pipe support for a LUMINAIRE.
spun. A light diffuser made of gauze.
spyder. See SPIDER DOLLY.
spy film. A feature motion picture, usually melodramatic, in which the protagonist engages in cloak-and-dagger conflict with foreign agents.
squawk box. A small loudspeaker, usually part of a public-address system.
squeegee. A wiping device used to eliminate moisture from film in the processor.
squeeze. To compress film images horizontally with the use of an ANAMORPHIC LENS.
squeeze ratio. The amount of horizontal compression in relation to the height of an image.
squib. A planned minuscule explosion used in an action shot.
squibs. Tiny explosive caps attached inside an actor's clothing, over protective metal plates, which can be exploded by remote control by wires that stretch from each cap (squib) to a battery; used to simulate exploding bullets.
stabilizer. A special CAMERA MOUNT used to control vibration.
stage. 1. A motion-picture studio, a SOUND STAGE. **2.** The television studio area in which a production is filmed.
stage brace. A triangular wooden support used to hold FLATS upright; a support strut for scenery.
stagehand. See GRIP.
stage left, stage right. The performer's movement as he/she faces the camera.
stag film. A film that emphasizes explicit sex, usually made to be shown to all-male audiences.
staging. The overall concept for a film, including action, lighting, scenery, costumes, and sets.
stand. An adjustable support for a LUMINAIRE, GOBO, REFLECTOR, etc.
standard. 35mm film.
standards. Specifications for film and equipment established by the AMERICAN NATIONAL STANDARDS INSTITUTE.
stand by. A warning to television performers and crew to be ready for filming.

stand-by. A performer or program ready to be used as a replacement if needed.

stand-in. A person who bears some physical resemblance to a lead performer and is used as a substitute during the time-consuming adjustment period for lights and equipment; sometimes filmed in scenes when the performer appears at a distance too far from the camera to be recognized.

standing set. A set that is used continuously in a television series: e.g., the garage in *Taxi.*

star. 1. A popular leading performer who appears only in principal roles. **2.** STAR FILTER.

star filter. A line-engraved lens filter that causes radiant beams (with a pointed-star effect) to extend from highlights in an action field.

star system. Filmdom's system of building a personality into a star performer and stressing his/her appearance in a film above all other creative aspects of the film, a system begun in the 1920s, that was continued through the 1960s, but today is on the wane.

start mark. A film FRAME indication that shows the initial position of the film to be used in starting any one of several editing procedures.

stat. See PHOTOSTAT.

state film commissions. Groups appointed in several states to attempt to attract film production companies to films in that state.

static. Unwanted snapping or crackling sound in an audio system.

station. A television (or radio) broadcasting facility that has a specific frequency for program transmission, authorized by the United States government.

stationary matte. A painted MATTE used in COMPOSITE photography to create an illusion that the action actually occurs at the site of the matte, which forms a background; the effect is accomplished through DOUBLE EXPOSURE.

station break. A brief pause, usually at half-hour intervals, in program transmission for call-letter identification of the television station, required by the Federal Communications Commission. See CALL LETTERS.

station identification (ID). A ten-second television commercial, of which two seconds are used to announce that particular station's channel number and call letters. See CALL LETTERS.

station manager. The executive who oversees the advertising, research, community affairs, and engineering departments of a television station; is also involved in other departments such as news, programming, and sales.

Steadicam. Trade name for a servo-controlled camera with a support that can be attached to the operator's body.

steadiness test. A test designed to ascertain the degree of play present in the film channel of a camera, done by photographing the same object from two slightly different angles on the same strip of film; when the film is projected the objects should remain steady in relation to each other.

step-contact printer. Contact printing in which the film being copied and the raw stock are advanced intermittently frame-by-frame and are exposed to the PRINTER LIGHT only when both are stationary. See CONTACT PRINT.

step lens. A condenser lens with concentric prisms on the plano (flat) side.

step on. To begin to speak before another performer has finished his/her line of dialog or comment.

step-optical printer. An OPTICAL PRINTER in which each frame is exposed individually.

step printer. A film laboratory machine that holds each FRAME stationary during exposure.

step printing. A printing procedure in which the film is printed one FRAME at a time.

step-prism, stepped-prism. A glass refraction device that contains prism wedges that vary slightly away from the center and help form a tight beam of light from a lamp filament.

step table, step wedge. See WEDGE.

stereo(phonic) sound, stereo. Sound reproduction in which two or more separated microphones are matched to separated playback speakers, thereby producing the effect of three-dimensional sound.

stereophonic variable-area track. A double VARIABLE-AREA SOUND TRACK on which the sound carried on each half of the track has been picked up by a separate microphone.

stereoscopic. See THREE-DIMENSIONAL.

stereotypes. Characters who have become familiar to audiences because of their frequent appearances in films: e.g., the bigoted, pot-bellied hick sheriff.

sticks, stick it. The cameraman's request for the CLAPSTICK operator to hold the slate in front of the camera and to clap the sticks together.

still. A single still photograph of a performer.

still background. A background in which there is no motion.

still frame. A single film or videotape FRAME held as a continuous shot.

still man. A photographer assigned to a film crew to take still photographs during production.

sting. 1. A single musical chord or note used as dramatic punctuation in a scene or shot. **2.** Occasionally refers to a single shout, also for dramatic effect.

stirrup. A device used to suspend a LUMINAIRE.

stock. 1. Raw film stock, unexposed film or videotape. **2.** STOCK FOOTAGE or STOCK SHOTS that are cataloged and stored in a film library for reuse.

stock character. A familiar character type, such as a bumbling absent-minded professor, or a performer who is identified with a stereotypical role.

stock company. A group of performers who play sustaining roles in a television series, either comedy or dramatic.

stock footage. Several feet of film containing STOCK SHOTS.

stock-footage library. A facility in which STOCK SHOTS and STOCK FOOTAGE are cataloged and stored, licensed for reuse.

stock music. See LIBRARY MUSIC.

stock part, stock role. A role calling for a STOCK CHARACTER.

stock shots. Shots that are stored in a motion-picture or television station library and used over and over again; they include famous places and buildings, scenic areas, historical events, etc. In television, stock shots include those used frequently during the lifespan of a particular series.

stock sound. See LIBRARY MUSIC.

stop. The lens aperture opening that controls the amount of light that enters the lens, calibrated from 1.5 to 22.

stop action. See STOP MOTION.

stop-action photography. A technique in which objects are shot one FRAME at a time at regular intervals so that projection at normal speed will show the changes that occur during a period of time: e.g., the growth of a puppy to a full-grown dog.

stop bath. A chemical solution, usually containing acetic acid, used to stop the action of the film developer in a FILM BATH.

stop down. To reduce the diameter of a lens aperture by adjusting the iris diaphragm.

stop frame. See FREEZE FRAME.

stop leader. A blank strip of film that indicates projection has been stopped on a single reel.

stop motion, stop action. 1. A special trick effect in which performers stop all motion during a shot, while some object is taken from or added to the action area, then resume action to the conclusion of the shot. The illusion is created that the performers or objects appear or disappear mysteriously. **2.** Time-lapse and single-frame cinematography, in which subjects move only slightly between exposures. See STOP-ACTION PHOTOGRAPHY.

stopwatch. A watch, calibrated in feet and frames, used to time film action and narration.

Storage Instantaneous Audimeter. See SIA.

story. A written (or occasionally verbal) condensation of a narrative from which a film script can be created.

storyboard. 1. Planned shots depicted through sketches or photographs with the dialog, music, and sound effects written in for each shot, used in the preparation of both live and animated films. **2.** The layout of the audio and video portions of prospective television commercials drawn on paper in separate frames.

story conference. A session between film writers and production personnel during which the merits of a script are discussed and suggestions are made for plot and/or characterization revisions.

story editor. A person who reads stories submitted to a film production company and recommends or rejects them as production material; occasionally edits a story before it is turned over to a scriptwriter.

story line. The plot outline of a film script. (Can be condensed to one or two sentences.)

straight cut. Two adjoining shots, cutting from one to the other, between which there is no intervening optical effect.

straight man. The actor against whom a comedian plays.

straight up. Indicates that the second hand on a clock is at 12.

Strauss lens. A soft-focus lens developed by Karl Strauss, the first Academy Award winner for cinematography (1927).

streaking. Distortion in a television picture which extends objects horizontally beyond their natural spheres.

streamer. A lateral mark made by a wax-base film-editing pencil on a WORKPRINT during projection, for editing purposes.

stretch, stretch out. To increase the length of a spoken narration by slower delivery, long pauses, etc.

strike. To dismantle and store props or equipment used on a set.

stringout. All workprinted camera shots joined together in the correct sequence for a final print. See ASSEMBLY.

strip. 1. A television program broadcast regularly at the same time on each weekday. **2.** A row of lights usually containing five 1,000-watt bulbs. **3.** To remove insulation from equipment.

stripe. A clear 35mm sprocketed film containing a magnetic oxide stripe for recording a single SOUND TRACK.

stripe filter. The vertical surface stripes on a camera pickup tube capable of breaking up an image light into red-blue-green components without use of DICHROIC MIRRORS.

striplight. A narrow box, open along one side, which contains a row of lamps and is used as a floodlight for CYCLORAMAS; usually contains 1,000-watt bulbs.

stripping. Repeating the broadcast presentation of several years of a long-running television series ACROSS THE BOARD.

strip-title. A title that moves horizontally across the screen.

strobing. A skipping or rotary motion of a filmed object at a speed inconsistent with the PERSISTENCE OF VISION phenomenon; can be caused by a PAN SHOT made too rapidly.

stroboscope. A device that emits light at regulated intervals.

structural film. Films in which unusual camera movements and photographic techniques are used: e.g., repeated LOOPS, extended ZOOMS, extremely slow camera movements, FLICKERS, and images photographed from the screen on which they are projected. A single film seldom contains more than one of these techniques.

structuralism. A study of cinema that uses BINARY OPPOSITION as a basis for its approach to the subject.

structure. The framework of a film developed through shots and scenes.

strut. A brace used to support scenery.

studio. 1. A sound-proofed room for production of motion pictures or television programs. **2.** A film production company.

studio camera. Any large camera, usually mounted and mobile, used on a studio set.

studio exteriors. 1. Sets that give the illusion of being outdoors but are built inside a studio. **2.** Shots made on such a set.

studio film. Any film that is shot entirely on a studio set or sets.

stunt check. Extra pay for hazardous work or for an extraordinary performance beyond the line of duty.

stunt man, stunt woman. An actor who performs the dangerous work in a film, usually doubling for the leading performers.

STV. See SUBSCRIPTION TELEVISION.

style. The particular manner in which a film is written, directed, performed, or produced.

subbing layer. A layer of adhesive that binds a film EMULSION to its base.

subjective camera. Referring to scenes shown from the point of view of the camera so that audience reaction is immediate and intensified.

subjective camera angle, subjective angle. A camera shot made from the point of view of the performer. Cf. OBJECTIVE CAMERA ANGLE.

subplot. A plot secondary to the principal plot but important to the structural creation of suspense.

subscriber. A television home viewer who pays for CABLE TELEVISION or SUBSCRIPTION TELEVISION transmission.

subscription television (STV). The broadcast form of pay television that is distributed by commercially licensed stations as an on-the-air signal (in contrast to a cable television service, which must be delivered by cable, wired street by street, house by house). The scrambled signal can be decoded only by a special device attached to the set for which the viewer pays a monthly fee; all outlets are independent UHF stations (channels 14 to 83).

substandard. See NARROW GAUGE (FILM).

subtractive primaries. The colors cyan (dark blue), magenta (purplish shade of red), and yellow.

subtractive process. A photographic means of viewing a particular scene and rearranging colors with three light filters, each of which represents a primary color. The process involves the making of a black-and-white negative of the subject through each filter. From each of these negatives a dye-positive image is made with the color of the dyeing complementary to the color of the filter used in each negative. The three negatives superimposed on each other result in a combined transparent positive image in natural color.

subtitle. An explanatory caption superimposed over the filmed action, shown at the bottom of the frame, usually a translation of a foreign-language sound track.

succès d'estime. A film that is acclaimed by critics but does not achieve popular success.

suction mount. See LIMPET MOUNT.

suds. SOAP OPERAS.

sun gun. A portable high-intensity LUMINAIRE.

sunk up. Referring to voices, music, or sound effects that are synchronized with the picture.

sunlight. Light from the sun; occasionally used to mean daylight generally even if the day is cloudy.

sunshade. A metal hood attached to the end of a lens to block out direct sun. See MATTE BOX.

sun spot. See PARABOLIC.

Super 8mm film, Super 8. An enlarged-frame version of standard 8mm film that has seventy-two frames per foot.

superimpose, super. To place one picture over another by using two different cameras or double printing.

superimposed titles, supered titles. The electronic addition of titles over a still shot or moving background images. In television, called *superimposition* or *super.*

Super 16. A 16mm film process in which the frame is increased to cover most of the sound-track area.

supply reel. See FEED REEL.

supply roll. The roll of raw film in a camera. See RAW STOCK.

supply spool. The spool in the camera on which the unexposed film is wound.

suppression. The reduction of aberrant frequencies to the required level.

surface, surface noise. Extraneous noise on a phonograph disk heard during recording or rerecording.

surrealist film. A motion picture that attempts to portray subconscious reality, usually with distorted or irrational effects.

surround sound. Music and other sound effects on a separate RELEASE PRINT track, played on loudspeakers placed at the sides of a theater so that the audience seems to be seated in the center of the sound.

survey. See LOCATION SCOUTING.

suspense. A plot technique used to keep the audience in a state of uncertainty or anxiety concerning the resolution of an exciting or harrowing situation.

suspension of disbelief. The audience's necessary acceptance of and belief in a dramatic situation, however bizarre or improbable, while they are viewing the drama.
sustainer. A local station or network program that does not have a sponsor.
sweep. 1. The electronic television picture tube scan. **2.** MILK SWEEP. **3.** SWEEPS.
sweeps. An intensified week-long audience rating contest among the major networks, held four times a year (in November, February, May, and July), during which each network presents schedules packed with the programs it believes to be its strongest contenders; intended to strengthen the network affiliates' bargaining power with advertisers.
sweetening. 1. The addition of new sound (singing) to a voice track. **2.** The addition of recorded applause or laughter to the actual audience reaction.
swing dissolve. Two swish PAN SHOTS connected by a DISSOLVE.
swinger. A FLAT that is swung out of the path of a camera DOLLY.
swing gang. Production crew members who work on the night shift.
swish pan. A rapid horizontal PAN SHOT in which the images are blurred; often used as a transitional shot between two locations.
switchback. A camera shot that returns to the original action from a CUTAWAY shot.
switcher. 1. The input control console used to select or mix the video output. **2.** TECHNICAL DIRECTOR.
symbolism. Any object or action in a film that is meant to suggest something else: e.g., a candle that goes out at an invalid's bedside suggests the coming of death.
sync, synchronism, synchronization. The precise aligning of the audio and picture elements of a film so that they coincide.
sync beep. See SYNC POP.
syncing dailies. Collecting the picture and sound WORKPRINTS of each day's shooting for synchronous projection.
sync generator. The electronic pulsing device that controls television picture SCANNING.
synchronization rights. Permission to use a recorded musical composition for animation background music.
synchronizer. A geared apparatus used to facilitate editing the film and sound track simultaneously.
synchronous motor. An AC (alternating current) motor with speed that is controlled by the frequency of the applied power source.
synchronous sound. Sound that is in SYNC with the picture.
synchronous speed. The same speed maintained by the camera, recorder, and projector during filming.
sync mark. An editor's mark placed on the working LEADER of a filmstrip to indicate the SYNCHRONIZATION point for other strips of film.
sync pop. A one-frame BEEP produced by a piece of quarter-inch magnetic

track attached to a magnetic film leader; used to provide a visual synchronization cue when synchronizing the optical track with the original film.

sync-pulse. 1. A camera timing device used to maintain recorder speed synchronism when the audio is played back. **2.** A minute voltage pulse introduced into a video signal during the BLANKING INTERVAL in order to provide exact synchronization between transmission and reception.

sync-pulse cable. A cable used to transmit SYNC-PULSE signals from camera to a tape recorder.

sync-pulse generator. A tiny camera-contained generator that transmits a SYNC-PULSE signal to be fed into a tape recorder.

sync-pulse system. Any recording system that places a synchronizing signal on magnetic tape to be used as a reference when the tape is played back so that the time reference will be the same as when originally recorded.

sync punch. A hole punched in a film SOUND TRACK or LEADER to be used as an audible cue when synchronized with other strips.

sync roll. The vertical rollover of a television picture, caused by some interference in the circuit.

sync signal. The voltage pulse that controls the scanning synchronization.

syndication. The production of a broadcast series to be marketed to independent stations; sold nationally on a market-by-market basis in contrast to network productions, which are distributed nationwide to a predetermined group of affiliated stations.

synopsis. 1. A very brief condensation of a film plot; can be written in two or three sentences. **2.** A summary of a film intended to be included in a film library catalog.

sweetening. The concluding mixing session on a film SOUND TRACK for the RELEASE PRINT.

T

tab. A foil-faced sensor in a film negative used in the timing of prints.

table top. Referring to close-up photography of small inanimate objects.

tachometer. A device that indicates the frame speed of a camera.

tag. 1. A short live announcement added at the conclusion of a recorded broadcast commercial. **2.** A brief scene that follows the climax of a film, used as a DENOUEMENT to wrap up the action.

tag line. A line, usually spoken by a comedian, which is the final and climactic line in a routine.

tail. The end of a reel of film.

tail-away shot. A shot of a performer moving in a direct path away from the camera.

tailgate. The film projector on an OPTICAL PRINTER.

tail leader. A LEADER used at the end of a strip of film or the end of a reel of tape.

tail out. Referring to a film or tape reel that has the end on the outside of the reel, requiring rewinding before projection or playback.

take. A camera shot of a single piece of action, usually identified on film by the SLATE or on a SOUND TRACK by voice.

take a 42. The announcement of the minimum meal break for a film production crew, forty-two minutes.

take board. See NUMBER BOARD.

take it. See PRINT IT.

take sheets. The script clerk's detailed records of daily activities on the set or location.

take-up. Referring to any mechanism such as a reel on which film or tape is rolled from a camera gate, projector, or editing device.

take-up plate. The horizontal TAKE-UP REEL on an editing table.

take-up reel, take-up spool. 1. The reel on which film is wound after it has passed through the projector. **2.** The spool on which exposed film is wound in a camera.

taking lens. The lens that is used to make a shot when there are two or more lenses on a TURRET.

talent. 1. Performers, usually professionals. **2.** The stars of a production. ("Put out a call for the talent.")

talent agency. An agency that represents performers in marketing their talents and in contract negotiations.

talent agent. A performer's agent. See AGENT.

talent scout. A studio employee who searches for talented performers among amateurs or unknown professionals.

talent union. A labor union to which professional performers belong. See AFTRA, SAG, and SEG.

talkback. A microphone/speaker communication system that operates between the television studio and control room.

talking heads. Referring to **1.** a medium-close shot of two persons in which there is a lot of dialog and little action, **2.** an interview in which two or more persons appear, usually a pejorative term.

talk show. A broadcast program in which a host or hostess introduces and chats with show-business personalities, national or international celebrities, and other persons currently in the news, usually before a studio audience.

tally light. The red light on a television camera which flashes when its particular shot is being transmitted.

T and A (tits and ass). Television program featuring beautiful young women and revealing as much of their anatomy as is allowable on the home screen.

tank. A large outdoor pool on a studio lot in which scenes on bodies of water, such as lakes, rivers, and oceans, are shot. (Alfred Hitchcock's *Lifeboat* was shot entirely in a tank.)

tank shot. A shot made in a tank in which MINIATURES are often used, though live action can be filmed under the proper conditions. See TANK.

tape. 1. A smooth plastic tape coated with metallic oxide, used for the electronic recording of sound or television picture patterns. **2.** To measure the distance between the focal plane of the camera to the object to be photographed in order to get the correct focus and PARALLAX. **3.** To record sound on magnetic tape.

tape guide. The grooved metal posts on each side of a MAGNETIC HEAD used to align the tape correctly.

tape hook. The hook to which the measuring tape is attached when measuring the distance from the camera focal plane to the subject of the shot to be photographed.

tape lifter. The metal arm that removes the audiotape from the record or playback when moving the tape fast forward or rewinding it.

tape recorder. A sound recorder, electronic or mechanical, that records magnetic sound on tape for playback.

tape speed. For audiotapes, 1⅞, 3¾, 7½, and 15 inches per second when played at normal speed.

tape splice. A film splice (BUTT SPLICE) in which the ends are joined by a short length of splicing tape and do not overlap.

tape splicer. A mechanism used to cut film so it can be spliced with no overlapping; thin SPLICING TAPE is used to join the ends together.

tapping the tape. Making musical beat indication marks on a moving sound track.

target. 1. A small metal disk used as a GOBO to block light from the camera lens or from portions of the set or wherever it is not wanted; also called a *dot*. **2.** See MOSAIC.

target audience. The particular audience with which a film is expected to find popularity: e.g., teenagers for surfing films.

TBA. A program or information *t*o *b*e *a*nnounced at a later date.

TBA (Television Bureau of Advertising). An organization of networks, stations, station sales representatives, and program producers/syndicators interested in the improvement and expansion of television as an advertising medium.

TBD. A course of action *t*o *b*e *d*etermined at an unspecified date.

T core. A plastic core, two inches in diameter, on which rolls of 16mm film can be wound up to 400 feet in length.

TD. See TECHNICAL DIRECTOR.

tearing. An interruption in the horizontal picture due to lack of synchronization in the electronic picture tube SCAN.

tearjerker. A film that has a heavy concentration of pathos.

teaser. A brief provocative scene that serves as a lead-in to a teleplay, often shown before or under the titles.

technical adviser. A person, usually not a member of the film industry, who is employed to provide authentic information for a specific film or television series: e.g., a police officer on a film about cops.

technical director. The engineer who operates the video control console.

Technicolor. A color-separation process in which color prints are made from three black and white negatives, one for each primary color (red, blue, and yellow) by using cyan, magenta, or yellow filters in the printer.

tee. A wood or metal device in the shape of a T, Y, or triangle, used to steady the legs of a TRIPOD.

Telco feed, Telco patch. A cable connection between a television station and the local telephone company.

telecast. A television broadcast.

telecine. The apparatus used to transmit feature motion-picture films on television.

teleconference. A television format that allows a person or persons in one location to be interviewed by reporters in various places across the country; the proceedings are beamed to a SATELLITE, transmitted to earth stations, and then to the locations where the reporters watch the subjects on a huge television screen and query them by telephone.

telephone coincidental interview. An audience survey technique in which persons are phoned and asked whether or not they are watching television and, if so, what they are watching.

telephone recall interview. An audience survey technique in which the recent television viewing habits of persons are inquired about by telephone.

telephoto distortion. The compression effect that occurs in shots made with a TELEPHOTO LENS: distant objects appear closer and action toward or away from the camera appears to be slowed down.

telephoto lens. An OBJECTIVE CAMERA lens with a great focal length and a narrow angle of view, used for making shots of distant objects or action so that they seem much closer.

Teleprompter. Trade name for a script-cueing mechanism that is placed near the camera and from which performers can read their lines as the script rolls vertically before them; lines may also be made visible by the use of a half-silvered mirror placed at a 45-degree angle so that performers seem to be looking directly into the camera.

telerecording (TVR). A filmed recording of a television broadcast made by photographing it directly from the CATHODE RAY TUBE of a television receiver, usually resulting in a film of inferior quality.

TELESAT. A *tele*vision *sat*ellite system launched by Canada in 1972.

teletext. A new video technology (still in the experimental stage), which will combine the written words of newspapers and magazines for showing on home television sets to which special equipment has been attached; will allow home viewers to receive written general, sports, and entertainment news, weather and traffic reports, stock-market information, and many other features.

telethon. An extremely long television entertainment program presented to raise funds for a particular charity or public-service organization.

television (TV). **1.** A process for the electronic transmission of pictures together with sound to receivers. **2.** The broadcasts themselves and/or the set on which they appear.

Television Bureau of Advertising. See TBA.

television commercial. See COMMERCIAL.

television cutoff. The cropping of a film frame due to faulty transmission.
television distributor. A company or corporation that produces and distributes programs for the SYNDICATION market.
television game. Referring to a microprocessor attachment that permits control of the CATHODE RAY TUBE (CRT) viewing area on the home receiver, used to play VIDEO GAMES.
television home. See HOUSEHOLD.
television mask (TV mask). 1. A mask used in a viewfinder or by an animation artist to indicate the boundaries of the SAFE-ACTION AREA on the television set. **2.** A mask used in designing titles so they will fall within the SAFE-ACTION AREA.
television receiver (TV receiver). A specially equipped box that is capable of sensing and receiving broadcast video signals, removing them from carrier frequencies and producing them as visual images on a CATHODE RAY TUBE (CRT).
televison safe-action area. See SAFE-ACTION AREA.
television set (TV set). See TELEVISION RECEIVER.
telly. English nickname for television.
Telstar. The original international television SATELLITE launched in 1962.
tempex. A customs form needed to transport film equipment from the United States for temporary use on locations in Europe.
template. An opaque sheet, placed inside a spotlight, that has been cut into a pattern in order to cast a patterned shadow.
tempo. The pace at which a motion picture moves.
ten. A broadcast advertisement that runs ten seconds.
tenner. A 10,000-watt spotlight with a Fresnel lens.
terahertz. One trillion HERTZ.
terawatt. One trillion WATTS.
terminal. The connection place for power equipment.
test. An exposed and processed strip of film used to judge whether or not it is acceptable for further printing.
test commercial. A broadcast commercial used to test audience reaction.
test film. Film used in various pieces of equipment to test operation performance.
test pattern. An optical chart used to determine the proficiency of the camera.
test strip. See TEST.
tetrode. An early amplifying vacuum tube that contained two variably charged mesh grids that regulated the electron flow between the negative filament and the positive plate.
TF (till forbid). Referring to a broadcast schedule having no initially specified cutoff date but cancelable at the request of the advertiser.
theatrical film. A feature film produced to be shown in motion-picture theaters.

theme. The principal idea or message of a motion picture.

theme music. Music used in motion pictures as background sound, heard frequently throughout the film, sometimes in association with a particular character or place.

theme song. 1. A song used to identify a particular television program. **2.** A song in the THEME MUSIC of a feature film or one that is prominently used throughout the film.

thermoplastic. Describing a technique of recording an image by an electron beam that deforms the surface of plastic film.

thin. Describing a negative that has not been exposed long enough to create a sharp image.

thirty. A broadcast advertisement that runs for thirty seconds.

35mm, 35mm film. Motion-picture film stock that is 35 millimeters wide with four perforations per frame on each side; has sixteen frames per foot.

35mm blow-up. A 35mm film made from film with a smaller gauge.

Thirty Rock. The nickname for the headquarters of NBC, located at *30 Rock*efeller Plaza, New York City.

thread, thread up. To place film or tape accurately in the path of a projector or other film mechanism for TAKE-UP.

threading path. The path followed by a filmstrip through film equipment; occasionally guided by a line or diagram.

three-camera technique. A term used to indicate that a film is shot with more than one camera, usually on the studio set.

three-color process. Any film process in which the three primary colors (red, blue, yellow) are used.

three-dimensional, 3-D. Referring to a technique by which the images and scenes of a film appear to have unusual depth as if seen off-screen.

threefold. Describing studio FLATS equipped with hinges in three places.

360-degree pan. A shot in which the camera makes a complete circle.

three-shot. A camera shot of three performers, usually from the waist up.

three-wall set. A set that contains three walls, two of which usually are parallel, with the fourth side open.

through-the-lens. Referring to a camera equipped with a VIEWFINDER that can focus through the camera lens without PARALLAX.

throw. 1. The distance from a film projector lens to the screen. **2.** The distance from the LUMINAIRES to the ACTION FIELD.

throwaway. Describing dialog lines that are underplayed by a performer.

ticket. An operating license for television engineers, issued by the Federal Communications Commission.

tie-down 1. A chain or cable used to anchor a TRIPOD to a stage screw. **2.** A TEE.

tie-in box. See SPIDER BOX.

tight. Describing **1.** a close camera shot in which the subjects occupy the entire frame; **2.** material that fills the limits of the broadcast time allotted for it.

tight gate. A camera GATE that exerts pressure on the film it holds.

tight shot. See TIGHT (def. 1).

tight winder, tight wind. A REWINDER attachment, such as a SPINDLE, ROLLER, or HUB, which winds film onto a core (usually plastic).

tilt, tilt shot. 1. A shot made with a camera as it is pivoted in a vertical line from a fixed position. **2.** The movement of the camera in such a shot.

tilt wedge. A device used to increase the TILT of the camera mount.

time-base corrector. The playback circuitry of a videotape recorder that produces excellent picture SWEEP synchronization.

time-base signal. A signal recorded on the edge of camera film to correspond with a signal on a magnetic recording; facilitates the SYNCHRONIZATION of film and sound WORKPRINTS. See SNYC.

time-base stability. The consistency of a servo signal in a videotape head drive. See SERVO MOTOR.

time buyer. A representative of an advertising agency who buys commercial broadcast time periods for clients.

time check. The synchronization of studio clocks.

time coding. A method of encoding the film so that the picture and the sound track will operate in SYNC.

time-day part. A daytime broadcasting period sold to advertisers.

timed workprint. A WORKPRINT that has been corrected for faulty exposures by various PRINTER LIGHTS.

time-lapse cinematography. Precisely timed camera shots, made at periodic intervals in single frames and projected at normal speed in order to record (in accelerated motion) a process usually invisible to the naked eye: e.g., a full-blown rose emerges from a tightly furled bud.

time-lapse cutaway shot, time-lapse shot. A shot in which the action jumps from the preceding shot to different action, quickly followed by the next shot in which the original action (or time) has been advanced.

time-lapse motor. A motor attached to a camera in order to make periodic single-frame exposures for TIME-LAPSE CINEMATOGRAPHY.

timer. A crew member who judges how much PRINTER LIGHT and color correction are needed on each film shot.

timing. 1. The meticulous process by which commentary is synchronized with a film, minute-by-minute, often second-by-second. **2.** The inspection procedure used by the TIMER to determine how much printer light and color correction are needed.

timing card. A card on which indentations are made in order to control light changes and color corrections in the film PRINTER.

timing tape. A paper tape used for the same purpose as the TIMING CARD.

tinny. Describing a thin sound that lacks low frequencies.
tint. To add color to a film with dye, used in the early silent films for special effect.
title. 1. The name of a film or television program. **2.** Any line of information that appears on the screen and is not part of a scene, including CREDIT, ROLL-UP, MAIN, and STRIP titles.
title card. A cardboard sheet or animation CEL on which an artwork title has been printed to be photographed.
title drum. A cylinder on which titles are rolled and will appear on film in an upward rolling motion.
title music. Music used while the titles are being shown.
title stand. A support for the camera and TITLE CARDS used when photographing the cards.
T-number. See T-STOP.
Todd A-O. A wide-screen process in which 65mm film was used; initiated by the late producer Michael Todd.
to length. Referring to a television program timed precisely to match the period allotted for it.
tone. 1. A pure color hue to which black or white has been added. **2.** An audio line-up signal of 1,000 HZ (see hertz).
tone control. An electronic circuit filter used to change high- and low-frequency responses.
tongue. 1. A mount for a camera DOLLY. **2.** To swivel the dolly mount to the right or left. **3.** The order to swing the mount: e.g., "Tongue right."
top hat. See HIGH HAT.
top lighting. Direct lighting that comes from above the performers.
top 100. The hundred principal television markets in the United States, cities in which all three major networks can be seen on local stations.
torque motor. A tension motor used to crank take-up spindles in various mechanisms such as the camera and the printer.
total audience plan. A spot announcement package intended to reach the largest weekly audience.
total audience rating. The number of television homes in which a minimum of six minutes of a single telecast were watched.
total survey area. See TSA.
tower. The antenna of a television station.
"to wow them." Referring to the intention to captivate an audience.
TPO. The signal power output of a television transmitter.
trace. CATHODE-RAY TUBE images made by a moving electronic stream.
track. 1. Film audiotape or videotape SOUND TRACK. **2.** The planks or rails on which a DOLLY runs. **3.** To make a tracking shot with the camera following a performer's movements.
tracking, tracking shot. A shot in which the performer's movements are followed by moving the camera along its axis or by moving the camera

with the action on tracks laid for this particular scene.

tracking finder. A RETICLE viewfinder on a camera used for tracking and photographing missiles.

track laying. Referring to the laying of DOLLY tracks by GRIPS.

tragedy. A drama in which the principal character or characters experience profound suffering and inner conflict, which ends for them in death or disaster.

trailer. Promotional shots from a coming attraction shown in a commercial theater before or after the FEATURE FILM; sometimes used on television to promote a forthcoming feature movie or a MINISERIES.

training film. A film, usually one that is part of a study program, used for instruction in skill-related tasks.

transcription. Any sound that is reproduced for broadcast, regulated by the code of the American Federation of Television and Radio Artists (AFTRA).

transducer. A device that converts electrical energy to mechanical or magnetic energy or vice versa: e.g., a microphone or loudspeaker in which the sound vibrations are changed into electronic pulses.

transfer. 1. To rerecord the sound from an original recording (or any subsequent recording). **2.** The rerecorded sound. **3.** The filmed copy of the image on a television screen.

transformer. A device used in the changing of voltage in an electrical system.

transient. A brief irregular signal that occurs during input change.

transistor. An infinitesimal SEMICONDUCTOR device that has greater control and amplification capabilities than the larger (now obsolete) vacuum tube; used in the reproduction of sound.

transit case. A wood-lined metal case used to protect reels of 35mm film when they are shipped.

transition. 1. Any optical means by which one scene is connected to or segued into the following scene, which differs in place and/or time (see SEGUE). **2.** Any means used for linking two shots. **3.** Sound, usually musical, that connects two sections of a television program.

translator. A low-powered, high-altitude station that receives broadcast signals and retransmits them on another frequency; used in areas surrounded by mountains or high hills where ordinary reception is not adequate.

translucent. Describing any material that transmits and breaks down the composition of light rays.

transmission. The passage of electromagnetic waves from the transmitting station to the receiving station.

transmission print. A completed tape or film approved for broadcast.

transmit. To send forth electronic signals from the transmitting station to the receiving station (television home sets) and thus actualize telecast programs.

transmitter. The equipment needed for the transmission of electronic signals.

transparency. A positive still image mounted on a glass or acetate support used for projection.

transparent. Describing light-transmitting material that does not alter the structure of the light rays and can therefore be seen through.

transponders. Receiving and transmitting units in a SATELLITE used in cable television broadcasts.

transport. Any mechanical device used to move tape past the recording and playback HEADS.

trapeze. A device used to mount set LUMINAIRES on overhead ropes or chains.

traveler. A stage set curtain that opens horizontally.

traveling matte. The combination within a PRINTER of an action shot with another action shot (or background) which has been shot previously. Used to achieve special effects in which an actor can "appear" in settings remote from the studio or can be made to "appear" in a dual role as he seems to perform in scenes with himself.

traveling matte printing. Film printing in which a TRAVELING MATTE is used.

traveling shot. A shot in which the camera is mounted on a dolly and moves with the walking subject of a shot; sometimes used interchangeably with TRACKING and TRUCKING SHOTS.

travelog. A short film depicting foreign lands, peoples, and customs, once popular for showing in conjunction with feature films but now usually made to be shown only to select audiences in special presentations.

treatment. A written summary of a film script that includes brief descriptions of the characters and the principal sequences around which the plot will be developed, as well as some action and dialog.

treble. A standard sound frequency range, from 3,500 to 10,000 HZ (see HERTZ).

treble roll-off. A gradual lessening of high frequencies.

tree. A high support stand on which LUMINAIRES can be mounted on horizontal bars.

triacetate base. A type of EMULSION coating for film.

triacetate film. Any photographic film with a TRIACETATE BASE.

triangle. **1.** A wood or metal framework used to hold the TRIPOD's three points steady as they are spread to maximum reach. **2.** The use of three kinds of lighting, KEY, BACK, and FILL, during filming.

triaxial. A lightweight cable used to convey sound, video, and control signals, in addition to power.

trickle charger. An AC converter used to drip DC electrons into a storage battery over a period of twelve to fourteen hours.

triggyback. Three twenty-second spot commercials, shown separately but sold for the price of a single one-minute spot commercial.

trim. To cut film in order to SPLICE it.

trim bin. A bin used by film editors to hold film that has been trimmed.

trims. The head and tail sections which have been removed from film shots.

tripack. Three-EMULSION color film on which the emulsions have been layered.

tripod. A camera support stand with three adjustable legs used to control height.

tripod head. A rotating camera mount designed to fit a TRIPOD.

trombone. A support clamp used to attach a set LUMINAIRE to a FLAT.

truck. The flanking movement of a camera DOLLY.

trucking shot. A shot made while the camera is moving along with the action, usually mounted on the flatbed of a truck or another conveyance.

TSA (total survey area). An audience research market classification of the American Research Bureau (ARB) that includes 98 percent of the total weekly home viewers.

T-stop. The lens setting that indicates the actual light transmission through the lens after absorption and reflection, unlike the F-STOP setting, which merely approximates the light transmission.

tubby. See BOOMY.

tube. See VACUUM TUBE. An obsolete glass-encased electronic device used in the control and amplification of sound, now replaced by the transistor.

tubesville. Describing an unsuccessful broadcast presentation or any proposed program considered unworthy of broadcasting.

tungsten. A lamp filament of 3,200 kilowatts.

tungsten-halogen, tungsten-halogen lamp. A small efficient lamp with TUNGSTEN encased in a quartz envelope filled with halogen gas.

tungsten index. See TUNGSTEN RATING.

tungsten lamp. Any lamp that contains TUNGSTEN.

tungsten rating. A number selected by a cameraman to indicate the relative sensitivity of a film to artificial light; sometimes suggested by the manufacturer of the film or by the laboratory that processes it.

turkey. An unsuccessful film.

turret. A rotary camera plate attachment that holds as many as five lenses, each of which can be moved quickly into place to facilitate rapid shooting; now usually replaced by the more recently developed ZOOM LENS.

turtle. A three-legged floor stand for the pipe support of a LUMINAIRE.

TV. See TELEVISION.

TVHH (television households). The number of HOUSEHOLDS in the United States that according to research estimates have one or more television sets.

TV-Q. An annual list issued by the Marketing Evaluation Company, which rates television stars according to their familiarity and popularity with the audience.

TVR. See TELERECORDING.

twang box. A sound-effects instrument with only one string of piano wire, used to produce a graduated *twang* sound.

tweak (up). To align electronic equipment correctly.

tweeter. A small loudspeaker usually used with a larger one to reproduce the high-frequency sounds.

twenty. A twenty-second broadcast commercial, usually containing eighteen seconds of sound.

two-bladed shutter. A projected shutter with two blades and two apertures, which reduce flicker when each film frame is projected twice.

twofold. Describing a FLAT that is hinged in the center.

two-reel comedy. An old silent film that was approximately 1,200 to 1,800 feet long and was contained on two reels.

two-shot. A camera shot of two persons in a frame, usually from the waist or shoulders up.

two-track stereo. Recorded and reproduced audio with left and right channel distribution.

two-wall set. A studio set with only two walls, which meet at an angle.

Tyler mount. A camera mount that eliminates vibration; used on helicopters.

typage. Selecting actors according to their physical type rather than their ability.

Type B. A kind of videotape that records each television frame in fifty-one short tracks.

Type C. A kind of videotape that records one complete television field during each Helican scan recording, used in FREEZE FRAME photography.

U

Uher. An extremely sensitive portable audiotape recorder used on location.

UHF (U). A secondary television broadcast band with an *u*ltra*h*igh *f*requency and narrow range (470 to 890 MEGAHERTZ), which includes channels 14 to 83.

ultrasonic. Having a frequency above the audible frequency range.

ultrasonic cleaner, ultrasonic film cleaner. A device that uses ultrahigh-frequency sound waves to remove dust or other particles from film.

ultraviolet. Just above violet in the spectrum, equivalent to light with wave lengths shorter than 4,000 ANGSTROM UNITS.

ultraviolet filter. See HAZE FILTER.

ultraviolet matte. A MATTE produced by photographing performers before a translucent screen that is back-lighted with ultraviolet light.

umbrella. An umbrella-shaped light-diffusing device made of white or silver-colored fabric.

unaffiliated. Describing an independent station that is not affiliated with a major network.

unbalanced. Referring to a film EMULSION that has been exposed to light, which does not maintain proper color temperature.

under. Describing any sound played softly in the background of a scene.

undercrank. To operate a camera at frame speeds slower than standard; produces unnaturally fast motion when projected at normal speed.

undercranked. Describing film that has been exposed when the camera was operating at frame speeds slower than standard.

underdevelop. To run film through a developer without giving it sufficient time to fully bring up the image.

underexposed. Referring to film that has received insufficient light in the camera, producing images without sharp definition.

underground film. A film independently produced that is usually experimental and contains subject matter not generally accepted for commercial films.

underground television. Programming provided by sources other than the establishment or network-sponsored and produced material.

underplay. To perform a role with restraint.

undershot. Describing exposed film that, due to an insufficient number of shots, is difficult to edit.

understudy. A performer who substitutes when necessary for a principal performer, having learned all dialog and action for the role.

underwater blimp, underwater housing. A watertight container from which underwater camera shots can be made.

undeveloped. Describing film that has not been immersed in a developer; unprocessed film.

unexposed. Describing **1.** film that has not been exposed to light, **2.** film that has not been used in a camera or printer.

unidirectional microphone. A microphone that picks up sound more easily from one direction than from others.

unilateral track. A SOUND TRACK produced with modulations generated on only one side of the track area.

unit. The production crew that works on a specific film.

United Scenic Artists. The trade union for scenic designers who work in motion-picture production.

unit manager, unit production manager. 1. The manager of a motion-picture production crew (unit) who is responsible for supplying whatever the crew needs, such as housing, meals, supplies, and transportation, in addition to taking care of the unit payroll. **2.** An employee of a network who is responsible for the coordination of a particular program's broadcast advertising material.

unit set. A studio set composed of furnishings and props that can be rearranged to achieve various effects.

Universal. The name commonly used for Universal Pictures, a major film production company.

universal leader. A film LEADER that contains identification and threading information marks needed by the projectionist to accurately thread the projector and make changeovers.

unmodulated track, unmodulated. The segment of a sound track on which there is no audio signal.

unscramble. See SCRAMBLE.

unsolicited script. A motion-picture or television script written ON SPECU-

LATION without prior agreement between writer and buyer.

up. Used in film scripts to indicate that the volume of background sound is to be increased: e.g., "Train whistle: in and up."

uplink. Electronic transmission from ground station to SATELLITE.

up shot. A shot made from a low angle with the camera tilted upward.

upside-down slate. A CLAPSTICK slate held upside down at the end of a shot to indicate that it is the end rather than the beginning, which is usually when the slate is shown.

upstage. 1. The rear area of a stage that is farthest from the camera or the audience. **2.** The act of a performer to deliberately divert the attention of the audience from other performers.

upstream. Movement of an electrical impulse from the subscriber or cable customer to the studio facility rather than the reverse, which is usual.

up the line. Movement toward the source of an electronic signal.

V

vacuum guide. A device used in a videotape recorder to hold the tape close to the record/playback HEADS.

vacuum tube. An early electronic control and amplification glass-encased device, made obsolete by the transistor. See TETRODE.

value. The measurement of the brightness of colors.

van. A truck that contains the necessary equipment for making videotape recordings from a location outside a studio.

vanda. Contraction of *v*ideo *and a*udio connection, used by telephone companies.

variable area, variable area sound track. An optical SOUND TRACK recording divided into opaque and transparent components along its length, achieving light transmission variations through the modulation widths.

variable density, variable density sound track. An optical SOUND TRACK recording with variable modulation density striations.

variable focus lens. A lens in which the focal length can be changed during shooting.

variable focus range. See ZOOM LENS.

variable hue sound recording. A sound recording with a photographic SOUND TRACK that has many color variations in contrast to those that have variations in monochrome density only.

variable-opening shutter, variable shutter. A cinematography camera shutter with a subsidiary leaf that can be set at different openings or can be entirely closed, used in FADE-INS and FADE-OUTS.

variable shutter control. A knob on a cinematography camera used to control the shutter diameter.

variable-speed control. A device in a camera or its motor that is used to change the speed of operation.

variable-speed motor. A cinematography camera attachment that controls the motion of the camera at various speeds.

varifocal lens. See ZOOM LENS.

vault. A fireproof facility with humidity and temperature controls for storing motion-picture film.

VCR. See VIDEO CASSETTE RECORDER.

velocitator. A kind of DOLLY that has a small crane for the camera and seats for the technicians.

velocity compensator. An accessory used on a video-tape playback to remove any aberration in the horizontal color transmission.

velveting sound. The act of cleaning a sound track by moving it through two pieces of soft dust-free cloth.

vertical resolution. The number of horizontal lines contained in a televised image.

vertical saturation. Extensive broadcast advertising aired during the daily broadcast hours in order to reach all members of a station's audience.

VFL. See ZOOM LENS.

VHF (V). The initial television broadcast band with a very high frequency (54 to 216 MEGAHERTZ), which includes channels 2 to 13. Compare UHF.

video. 1. The visual portion of a signal containing both picture and audio, transmitted via television (from the Latin "to see"). **2.** Early slang for television or anything seen on the home screen.

video analyzer. An electro-optical device used to SCAN a color negative in order to find the correct printing exposures.

video cassette. A container that holds a LOOP of half-inch- or quarter-inch-video tape on which both visuals and sound can be recorded; has two reels connected by the running tape in the MODULE, supply, and take-up. All major film studios provide some videotaped movies for home consumption.

video cassette recorder. A device that is plugged into home television sets to record or play prerecorded video cassettes.

videodisc. 1. A disk on which television recordings, both visual and audio, are made either with a laser beam, a stylus and grooved disk, or a stylus and grooveless disk. **2.** Equipment needed in slow-motion or freeze-frame effects.

videodisc player. A device that is plugged into home television sets in order to play (but not record) audio and visual material from VIDEODISCS.

video games. Games that can be played on home television sets with the use of a microprocessor attachment, which permits control of the CATHODE RAY TUBE (CRT) viewing area.

video leader. A standard HEAD and/or TAIL LEADER for video film approved by the Society of Motion Picture and Television Engineers (SMPTE).

video looping. Feeding video signals into several monitors simultaneously.

Videoplayer. A specially designed device capable of projecting SUPER 8 FILM into home television receivers.

video processing amplifier. See PROC AMP.

video signal. The electrical frequencies of a television image expanding from zero to approximately 4 MEGAHERTZ.

videotape. Flexible plastic 35mm tape coated with magnetizable iron oxide and used to record both sound and images for television broadcasts.

videotape recorder. See VTR.

Videovoice. A device that is capable of transmitting slow-scanned still images or freeze-frame television images over standard telephone circuits.

vidicon. A long-lasting television camera pickup tube.

viewer. 1. A mechanical/optical film-editing device that enlarges the filmed images and through which the film can be pulled by the editor at any desired speed. **2.** A person who watches television.

viewers per set. See VPS.

viewfinder. Any optical system that is part of a camera and is used to provide an exact (usually) magnified image formed by the lens as it is being recorded on the film.

viewfinder mismatch. Referring to use of a VIEWFINDER OBJECTIVE LENS or mask that is not adjusted to the lens of the camera being operated.

viewfinder objective, viewfinder objective lens. The lens of an optical monitoring VIEWFINDER, in which the image is formed.

viewing. Referring to videotape approved for broadcasting.

viewing filter. A filter used by a cameraman to view the lighting, either on the set or on location, in order to determine the degree of lighting needed to achieve satisfactory results.

vignette. 1. To cut the outer edges of a picture without reducing the size of the central image. **2.** The vignetted picture itself. **3.** A shot made through an opaque mask placed before a camera in which only the diffused center portion is visible.

virgin. Describing videotape on which no electromagnetic signal has been recorded.

visc. A video disk developed for use by home viewers.

viscous processing. A film-processing method of immersing film in chemical baths in which the liquid moves slowly over and around the film.

Visnews. An international organization that specializes in gathering news for television broadcasts.

visual. Any action or object that is seen on film.

visual effects. Special effects shown in motion pictures, often made during the printing process, unlike MECHANICAL SPECIAL EFFECTS, which are usually made by the camera.

visual primary. A film image that is more important to the scene or shot than the sound is.

vizmo. A device used in the BACK PROJECTION of backgrounds inserted into a live broadcast.

voice over (VO). The voice of an unseen actor or narrator heard during a scene or shot, used for narration, commentary, and inner monologs.

voice test. An oral audition by an actor or narrator to test the voice quality.

volume. The intensity of sound.

volume control. See FADER.

volume units meter. See VU METER.

VPS (viewers per set). An audience survey that counts number of viewers in the same television home.

VTR (videotape recorder). A device used to record both the television image and sound on magnetic tape in order to achieve immediate PLAYBACK.

VTR operator. The television engineer who is responsible for the operation of the videotape recorder in a studio. See VTR.

VU meter (volume units meter). A sound recorder/playback meter that indicates amplitude variations, either by calibrated decibels or by a percentage scale.

W

wagon wheel effect. An optical illusion seen on the screen when the spoked wheels of a moving vehicle seem to move counterclockwise; caused by the opening of the camera shutter at times when the spokes are in a retarded position instead of in a forward position.

walk-on. 1. A performer who has no lines to speak. **2.** The performance itself.

walk-through. A rough rehearsal during which no cameras are operating and usually no dialog is spoken as the performers move through the actions required in the shot.

walla-walla. The murmuring noise in a crowd scene in which no individual voices can be distinguished.

wall bracket. A fixed attachment used to hold a LUMINAIRE on a wall or FLAT.

wall plate. A kind of flat or wall-mounted pedestal support for a LUMINAIRE.

wall rack. An editing bin that is mounted on a wall.

wall sled. A metal bracket mounted on a flat or wall as a LUMINAIRE support.

wardrobe. 1. Costumes worn by performers. **2.** The studio department in which the costumes are made and stored.

wardrobe master/mistress. A crew member who obtains and cares for the clothing worn by performers in a film.

wardrobe truck. A truck equipped to hold and repair performers' clothing and to transport it between the storage facility and the set.

war film. A film made with a war background as an important component in the plot.

warm. Describing a television picture image that is slightly reddish or yellowish in tone.

warm-up. A brief live introduction used to entertain a television studio audience before the feature performance goes on the air.

wash. 1. To run developed and fixed film through a clear water BATH. **2.** The bath itself.

waste circulation. Referring to the broadcast of commercials to an audience unlikely to be customers for the product advertised.

water bag. A heavy rubber bag filled with water, used to act as a weight on the legs of various pieces of equipment on the set.

waterspots. Undesirable splotches created by a faulty drying procedure used on processed film.

watt. An electrical power unit equal to one ampere of current under one volt of pressure.

wave-form monitor. An oscilloscope tube used in the analysis and adjustment of signal characteristics. See SCOPE.

wave generator. See WAVE MACHINE.

wave guide. A straight metal pipe filled with nitrogen which can transmit several ultrahigh-frequency signals simultaneously.

wave length. The distance over which one cycle of mechanical or radiant energy travels; broadcast frequency.

wave machine. Steel rollers that revolve and move up and down, driven by an electrical motor; used in a large water tank to simulate the churning of waves for special-effects shots.

wax. 1. The wax solution applied to release prints. **2.** The waxing process.

wax pencil (crayon). A pencil with a wax base used in marking films.

weave. The undesirable sideways motion of film in a projector GATE.

wedge. A brief length of negative film in which each frame is made progressively darker to aid in laboratory testing.

weenie. See GIMMICK; MACGUFFIN.

Westar, Westar 2. Two geosynchronous United States communication satellites, launched by Western Union, used by Public Broadcasting System and pay cable systems.

Western film (movie), Western. A film with a locale in the western part of the United States, usually with a nineteenth-century rural, ranch, or small-town setting.

"Westinghouse Rule." The Prime Time Access Rule (the result of a petition by Westinghouse Broadcasting Company to the Federal Communications Commission), which reduces to three hours per night the time during which the top fifty market affiliates can broadcast network programs, forcing them to use more local programming.

wet cell. A storage battery that uses water to electrically produce a chemical reaction.

wet gate, wet gate printing. A printing process in which a coating of tetrachlorethylene solution is placed on the film prior to exposure; this fills any surface scratches and thus eliminates defects on the printed film.

wheeled tee, wheeled tie-down, wheeled triangle. A TEE that is mounted on wheels in order to achieve mobility for the tripod-mounted camera.

whip. An abrupt camera movement.

whip pan. See SWISH PAN.

wide-angle distortion, wide-angle effect. A pronounced foreshortening of subjects or objects when shot by a wide-angle lens in a camera placed close to them.

wide-angle lens. A lens with a short focal length and a viewing angle of more than 45 degrees.

wide-angle shot. A shot for which a WIDE-ANGLE LENS is used in order to reveal more of an action area than would be shown by a camera with a normal lens from the same distance.

widen. To draw the camera back, by DOLLY, from a position close to the subject; also to ZOOM back from the subject for the same effect.

wide open. Describing a lens that is set at its lowest F-STOP rating so the IRIS is opened as wide as possible.

wide-screen. Describing a picture-and-screen format in which the width-to-height ratio between the picture and the screen is wider than the standard 1:1.33 ratio, usually 1:1.65.

wide-screen release print. A RELEASE PRINT that has been reduced to conform to the wide-screen width-to-height ratio.

wigwag. A moving semaphore signal placed on a barrier to warn that filming is taking place in that area; semaphores sometimes have flashing red lights.

wild. Describing **1.** any device used in filmmaking in which precision cannot be maintained, such as variable-speed motors; **2.** the parts of a set that can be moved at will; **3.** separate recordings made of related material.

wild motor. See VARIABLE-SPEED MOTOR. A motor which has variable speeds.

wild picture. A motion picture not made on a synchronous sound track; the synchronized sound is added later.

wild recording. A recording made without camera synchronism.

wild spot. A broadcast commercial by an advertiser with national or regional sales which is aired during a local station break.

wild track. A SOUND TRACK made without camera synchronism.

wind. **1.** To roll film on a spool, reel, or core. **2.** To tighten a camera's spring mechanism by turning a crank or key.

wind gag. A protective device for a microphone to protect it from direct wind and its sound.

wind machine. A large fan used to create the effect of blowing wind during an action shot; occasionally airplane propellers are used.

wind up. 1. A hand signal given to performers to conclude their presentation when time is running out. **2.** WIND.

wing it. To perform "cold" without previous rehearsal; to IMPROVISE: e.g., without rehearsal, an actor will "wing it."

wipe. 1. An optical effect used as a transition between two shots in which the first often appears to be pushed vertically off the screen to the right by the gradual appearance of the second shot from the left, with a dividing line moving between them; other variations include the horizontal, diagonal, iris, and spiral wipes. **2.** To erase magnetic recordings.

wipe-off animation. See SCRATCH-OFF.

wireless microphone, wireless mike. A small cableless concealed transmitter used by a performer to broadcast a voice signal over a short distance to the receiver/recorder.

wire recorder. A magnetic recording device used prior to the development of the AUDIOTAPE recorder.

wire service. A broadcast news service offered by national or international press associations, delivered to the stations by teletype machines.

wire-tripping. The practice of stretching taut wires in the paths of running horses in order to make them fall.

Women in Films. A supportive organization for women in the motion-picture and television industries, acting as a clearing house for information on qualified women directors, producers, writers, etc.; maintains a foundation to give financial assistance to women for creative work in various fields.

Women in the Motion Picture Industry, International. A federation of clubs of women employed in the production, distribution, and exhibition of motion-picture and television films.

"woof." Technician's synonym for "Okay."

woofer. A loudspeaker in which a component carries low frequencies.

working distance. The maximum distance a performer can stand from a microphone for adequate sound pickup.

working leader. The blank end of film, used to thread rolls through editing equipment.

working title. The tentative title by which a film is known during production if a final acceptable title has not been chosen.

worklight. Any light used on a studio set for illumination before the set has been lighted for actual filming.

work picture. A filmed sequence put together by the film editor to determine its suitability, usually combined with the WORK TRACK.

workprint. 1. Any positive duplicate picture or sound track print used by the film editor during the editing process, in which, usually, all scenes are placed in sequential order. **2.** A rough cut of sound and picture combined made to preserve the original from damage.

work track. An audio sequence put together by the film editor to determine its suitability; usually combined with the WORK PICTURE.

worldize. To play back and rerecord live background sound on a location away from the studio.

wow. A periodic sound distortion caused by repetitive variations in the speed of a mechanical component in a sound system.

wrap. The successful completion of a shot ("It's a wrap").

writer. See SCREENWRITER, SCRIPTWRITER.

Writers Guild of America (WGA). The professional screenwriters' union, which has two affiliates: WGA-East (headquartered in New York City) and WGA-West (headquartered in Los Angeles).

writer's representative. See AGENT.

writing speed. The speed of contact between the videotape recorder head and the surface of the tape.

wrong-reading. Referring to the reversal from left to right on the screen of any written material, graphics, or images.

X

xenon lamp, xenon. A DC quartz-glass projection lamp that uses xenon gas (a colorless inert gaseous element) to provide durable and efficient illumination having a constant COLOR TEMPERATURE.

X-rated. 1. A motion picture defined in the Code and Rating Administration rating board of the Motion Picture Association of America as one unsuitable for persons under seventeen; the films do not carry the MPAA seal. **2.** The label given to pornographic films by the general public.

x-ray. A LUMINAIRE strip that hangs overhead on the set.

X-sheet. Written directions provided for the exposure of animation film.

Y

Y signal. A color luminance signal (4.5 MHz).

Z

Z core. A plastic core, three inches in diameter, on which is wound rolls of 16mm film more than 400 feet long.

zeppelin, zeppelin windscreen. A lengthened perforated tube attached to a SHOTGUN microphone to minimize noise from the wind.

zero cutting. A negative editing technique in which specially prepared A and B ROLLS are used to conceal splices made in the original and to preserve the original uncut.

Z.I. See ZOOM IN.

zip pan. See SWISH PAN.

Zoetrope. An early animation device that contained a pedestal-based revolving drum slitted so that, as it was spun around, the small figures in various positions on a strip of paper inside the drum seemed to be in motion when viewed through the slits.

zoom. 1. To alter the size of the action area from wide-angle to close shot without moving the camera, by means of a ZOOM LENS. **2.** A shot made with such a lens.

zoom back (ZB). See ZOOM OUT.

zoom blimp. A soundproof cover for a ZOOM LENS, used to blanket the whirring noise from the camera.

zoom drive. An electrical device that drives a ZOOM LENS, used to provide smoother action than when hand-operated by the cameraman.

zoom finder. A PARALLAX-type viewfinder to see the action area covered by a ZOOM LENS; made obsolete by the invention of the REFLEX VIEWFINDER.

zoom in (ZI). Instruction to increase the focal length of the ZOOM LENS during shooting in order to magnify the image as the size of the action area decreases.

zoom lens. A lens with a variable focal length that is effective both in eliminating lens changes and in simulating an effect of camera movement (rapid or slow) toward or away from a subject.

zoom out (ZO). Instruction to decrease the focal length of the lens during shooting in order to decrease the size of the image as the action area is magnified.

zoom shot. A shot made with a ZOOM LENS, which is changed in focal length while the shot is in progress.

Zoopraxiscope. An early motion-picture projector in which the photographic images were on a circular glass plate instead of film.